SEVEN MIRACLES

Performed by God So You Could Exist

John L. Leonard

Copyright

www.southernprose.com

Published by Each Voice Publishing
www.eachvoicepub.com

ISBN: 979-8-9863102-5-1

Dedication

This book is dedicated to the glory of Yahweh, the Creator God who performed the seven miracles this book enumerates.

Acknowledgments

The author would like to thank the following individuals: Lisa Leonard for her editing and cover art contributions, Wilfred Butler for being my primary admin for *The God Conclusion* Facebook page, Richard Suttles for the inspiration to write this book, and Mark Martone and the individual who uses the moniker Think Much? on Facebook, for their contributions in the manuscript appraisal process.

About the book

Two people deserve credit for inspiring this book. The first to spark ideas for the concept of *Seven Miracles* was Joe Rogan. In an interview with Stephen Meyer (Rogan has made similar comments to other guests), Joe said, "Did you ever read any Terrence McKenna? Terrence McKenna had a very funny thing that he said about science. He said science wants you to believe that it's all about measurement and reason IF you allow them one miracle. That one miracle is the Big Bang. That all things come from the most preposterous idea ever, that everything came from nothing in one big miracle."[1]

The second person who inspired this book is my friend Richard Suttles, who now calls himself The Fallen Ape-Theist on Facebook (Meta) but previously used the moniker The Faithiest Atheist. I would joke that Richard knows more Christian evangelists and apologists than I do, except that isn't a joke. Honestly, he does know them—I've seen photos of Richard having dinner with some of my favorite Christians, and frankly, I'm jealous. During our conversation on a podcast where we were discussing my book *The God Conclusion*, I jokingly suggested that I should call my next book *Three Miracles,* with the other two miracles being the origin of life and the rapid expansion of the early universe (also known as cosmic inflation) adding up to a grand total of three miracles, a number I personally liked because of the parallel to the three persons of the Trinity. However, Richard took my joking suggestion seriously and instead suggested the title *Four Miracles*, adding the miracle of consciousness. Richard is a very smart guy, and he's absolutely right—consciousness IS a miracle.

But why stop at four miracles? Perhaps the better question is, how did I go from four miracles to seven? Glen Scrivener, a minister and evangelist with the Church of England, also says

---

[1] Rogan, Joe. "If you allow them one big miracle Joe Rogan—Stephen Meyer", Jesus of America, April 18, 2025, https://www.youtube.com/watch?v=8A5raMfrZ4Q

there are four miracles that most people will accept. His first miracle is the same as Rogan's, that everything has come from nothing. Scrivener then adds order from chaos, life from inanimate, non-living matter, and consciousness emerging from mindless matter. Tough to dispute. Scrivener describes these four miracles as "Easter" miracles, meaning they involve life from the dead powers of gargantuan proportions.[2] In fact, all four of Scrivener's enumerated miracles have been incorporated in whole or as part of my full list of seven miracles. The origins of life and consciousness directly map to miracles in my list, and "order from chaos" describes everything from cosmic inflation to the diversity of living organisms.

Cliffe Knechtle might be my favorite Christian apologist. Cliffe is famous for visiting college campuses around the U.S. and evangelizing to students now with his son, Stuart. Recently, Cliffe identified **six** miracles that even atheists must believe, including:

1. Existence comes from non-existence.
2. Order comes from chaos.
3. Life comes from non-life.
4. The personal comes from the non-personal.
5. Reason comes from non-reason.
6. Morality comes from matter.

[2] Scrivener, Glen. "Persuading Thomas: How can modern doubters believe that Jesus rose? Revealed in the stars", Page 20. Evangelical magazine. July/August 2018.
https://www.evangelicalmagazine.com/article/persuading-thomas/

I've got to admit that Cliffe's list is excellent.[3] There's nothing wrong with his list of miracles.

Because I was working on this book before I discovered Cliffe's list and I'd already come up with my own list of miracles, I'm going to stick with my own list. Everything on Cliffe's list is represented on my list—order coming from chaos is represented as the fine-tuning of the universe, and morality is a function of consciousness. I could have limited my list to six miracles without adding the miracle of the individual, but I liked the evidence, and I strongly prefer the number seven to six, quite possibly for superstitious reasons, since I was born on the seventh day in the seventh month of the year. In the Bible, seven is the number of perfection and completion.

Seven has always been my lucky number (although this will be my ninth book. No one has ever accused me of being a genius when it comes to mathematics.)

---

[3] Unknown. "6 Miracles Atheists Believe", questionsintheology.com, May 12, 2024, https://questionsintheology.com/6-miracles-atheists-believe/

TABLE OF CONTENTS

SEVEN MIRACLES

INTRODUCTION

The definition of a miracle, according to the American Heritage Dictionary, is:

1. An event that appears inexplicable by the laws of nature and so is held to be supernatural in origin or an act of God.

2. One that excites admiring awe. A wonderful or amazing event, act, person, or thing.

3. Merriam-Webster dictionary has a slightly different definition:

4. An extraordinary event manifesting divine intervention in human affairs.

5. An extremely outstanding or unusual event, thing, or accomplishment.

Yet in both instances, miracles are perceived to be the result of divine intervention or an act of God. As a result, modern medicine does not use the phrase "medical miracle" to describe a patient's miraculous recovery from an incurable illness. Instead, the medical community calls it "spontaneous remission." The medical community is not denying that a miraculous cure has occurred, but it is denying that a divine entity was involved. Philosopher David Hume described miracles as "violations of the laws of nature." But the miracles listed in this book are violations of nature *known to have occurred.*

This book defines a miracle as an act of divine intelligence, a.k.a. God. Miracles are events that are so unlikely they defy the laws of logic and probability. It is only because we have clear and incontrovertible evidence that some unbelievable event has occurred that we can believe the event is even possible. One example that readily comes to mind is the origin of life. If you simply listen to what the honest experts on the subject of abiogenesis have said, you could be led to believe that inanimate, inorganic matter will never become a living organism, no matter how long the problem has been given to solve itself. But we should never say never. At most, we might say "extremely unlikely." Even the most optimistic experts who think the mystery of abiogenesis might be solved one day still don't think the answers will be coming anytime soon. Researchers haven't come close to replicating the feat of animating lifeless matter. They don't even know how the components of a cell came into existence.

Words like *impossible* have sometimes been used to describe the problem, even though in statistics theory, "impossible" represents one of the two extremes on the probability spectrum and a certitude that doesn't lend itself to science very well. If and when the word impossible is used, it doesn't mean literally impossible. It means virtually impossible. To claim that one knows God performed a miracle would mean that one has literally eliminated every other potential explanation.

Sometimes, stupid good luck does produce an outcome. Did God perform a miracle in 1980, so the U.S. hockey team won a gold medal and defeated the unbeatable Soviet team, or did the U.S. team play the game of their collective lives while the Russians were caught off guard and underestimated the resolve of the American team? A movie about the game was even called

Miracle, but Herb Brooks, not God, deserves the lion's share of the credit for possibly the greatest upset in the history of sports. Personally, I doubt God had any interest in the outcome of the game. To be perfectly honest, God probably doesn't care one bit about football or hockey.

On the other hand, we have plentiful evidence on Earth that the problem of existence, which Herb Brooks could never solve even if he had fourteen billion years, has nevertheless been solved. All you need for evidence is an episode of *Planet Earth* to see that life on Earth is ubiquitous. There are millions of unique life forms in a wide variety of sizes and shapes, from elephants to mosquitoes. We have everything from microscopic-sized extremophiles to giant redwood trees. Life exists. It is an indisputable fact. More important to understand—life cannot evolve *until* it exists. Before evolution can ever become possible, abiogenesis must have already occurred. This means that if abiogenesis doesn't occur and cannot be explained by purely natural processes, there is no reason to believe evolution is true. No matter how persuasively the foundation for Darwin's theory can be argued using evidence and peer pressure, and no matter how good the alleged evidence for evolution is claimed to be, the theory of evolution utterly depends on the success of a hypothesis we have no reason to believe is even possible except for the telltale evidence of life itself.

You're supposed to never say never in science. Scientific theories never become proofs. New and better evidence can completely change and reshape the public's consensus opinion. However, this standard apparently does not apply to evolutionary biology. Beliefs about evolution simply cannot be challenged or even questioned without the risk of public ridicule and having one's character verbally assaulted. Darwinism has become comparable to the most zealous forms of religion, and its advocates do not tolerate questions raised about the theory. There is even an unholy alliance between the secular (atheist) advocates for evolution and those Christians who call themselves theistic evolutionists. In my opinion, the combination of disparate ideas is nonsense. God, the creator or intelligent designer, is a planner. Evolution, as described in textbooks, is a completely unplanned and undirected process. It's like trying to mix oil and water. The two ideas don't blend well.

According to evolutionary theory, humans are alleged to have descended from ape-like ancestors. Humans are clearly more advanced and intelligent than apes. So, the question becomes, can an unplanned and undirected process produce intelligence and explain improvements in physical design? Can a DNA blueprint even be called a design, or is it merely the illusion of design? If increased intelligence in humans is a random byproduct of unplanned and undirected processes, why should we believe this superior intellect is trustworthy, if we assume our thoughts are produced by chemical reactions in our brain as it codes for protein? In my opinion, the theory of evolution is incompatible with belief in God. I won't believe anything just because someone insists that I should believe it, mainly because I don't care what other people think. It's more important that I think. I care a lot more about the truth than I care about what other people might think of me.

If God exists, then evolution did not occur. God is part of the planned universe model, and evolution is allegedly unplanned. Either creation or intelligent design must have happened instead of theistic evolution. Evolution is part of the binary worldview that excludes God. That worldview turns out to be logically inconsistent and ultimately requires far more faith than theism. That is my claim. This book will explain how my claim is probably true.

AUTHOR'S QUALIFICATIONS

The only real qualification for becoming a professional writer is the ability to string together coherent sentences to communicate lucid and coherent ideas using the English language. That and the ability to find a good publisher. That last part was easy—I'm married to the editor-in-chief of Each Voice Publishing. In other words, I sleep with the boss. My first book, *Divine Evolution*, was initially published by a small publisher called ePress Online, a company that went out of business after the editor-in-chief and her successor both died within weeks of each other. The rights to my book were returned to me, so my wife took the finished manuscript and re-published it as the first title under Each Voice Publishing. If you want to say that all my books except the first are self-published, I won't object, because it's true.

Writing is not all that difficult, to be perfectly honest. However, producing books that are well-written and that people will want to read is quite difficult. Selling a book can be even harder. Most people would rather wait until the book gets made into a movie or comes out on video. In addition to writing books (in English), I am proficient in several computer programming languages. For many years, that was how I earned the majority of my income. While I will be writing about physics, chemistry, and biology, I need to stress that I do not consider myself an expert in any of those scientific disciplines and only claim to be a consumer of existential science.

I'm not a science geek. My field of professional expertise is limited to information and technology, as a former software developer. In computers, data is presented as a binary choice. Any bit of data can only be either a zero or a one. There are no additional rules about preceding or subsequent bits beyond the fact that bits are organized into groups called bytes. There are eight bits in one byte of information, which means you can create every character and number in the alphabet by stringing together different combinations of zeros and ones in these eight-bit packets of data. For example, in a computer, the number 0 is represented as 00000000. Simple enough. The number 1 is 00000001, but the number 2 is 00000010, and the number 3 is 00000011 according to the ASCII table. The number 7 is 00000111, and so forth. An uppercase letter "S" is

01010011, but a lowercase "s" is 01110011. It might look complicated, but it really isn't—you only need to look at an ASCII table to find these values yourself. The only real trick is knowing where to look and which table contains the information.

Of course, I didn't get paid for knowing that when I press the lowercase letter "a" on my keyboard, the computer generates 011000001, or that when the alphanumeric character 7 is pressed, 00110111 is conveyed. I got paid for *knowing what to do* with the "a" or the number 7 when that key was pressed. I was paid to help the computer accept raw data as input and produce intelligent output. I wrote software, and then after the software was sold to customers, I provided technical support for the product I'd helped write. Not to brag, but I was rather adept at my job. I traveled to Singapore, Australia, and Hong Kong on an expense account, which I doubt the company would have insisted upon if they thought I was a bumbling idiot.

Even so, I'm not throwing out a bunch of technical jargon and buzzwords to impress you. It is to demonstrate that I do know how computers work. I know less about DNA, but I do know DNA is easily twice as complex as machine language and more restricted. Instead of two choices (zero or one), there are four: adenine, thymine, cytosine, and guanine, but these nucleotides must be paired specifically. There are strict rules about the nucleotide's ability to pair with each other—adenine always pairs with thymine, and never with cytosine or guanine. In machine language, the only rule is that each bit must be zero or one and followed by another zero or one. The stream of binary data is organized into bytes of information.

The rules and restrictions for sequencing DNA are far more complex, even though machine language is definitely produced by planned and intentional processes, but DNA is allegedly created by unplanned processes. It is an assumption that unplanned and undirected processes could have possibly created DNA, but not a very logical assumption. Because DNA tends to break down quickly, we should not assume there was an infinite amount of time for DNA to form. The blind forces assumed to produce a successful undirected universe must work faster than entropy or decay can break down whatever is being built.

A miracle, by its definition, is a unique event, statistically speaking. Yet some people might argue that simply representing a statistical improbability with a near-zero chance of success is an irrelevant consideration if an event only needs to happen once in the course of the history of the universe. As long as an event can be said to be theoretically possible, no matter how preposterous it seems, it inevitably becomes a near certainty if only enough Time can be said to elapse.

Indeed, Time is the saving grace for the secular mind. Time and Luck are the gods of methodological naturalism. With enough Time and enough Luck, literally anything becomes possible. Perhaps monkeys could even learn how to shoot fireballs out of their buttocks,[4] simply if given enough time. While that last claim was meant to be taken tongue-in-cheek, there is a *Braveheart* parody video suggesting the phenomenon is theoretically possible.

SCIENCE, MATHEMATICS, STATISTICS, LOGIC, AND PROBABILITY

Science is a search for knowledge about our universe and the natural world that begins with an idea, called a hypothesis. Evidence is observed and collected until deemed sufficient to move the idea from untested hypothesis to scientific theory. Science is never settled. In science, there is evidence, but never proof. The evidence for a scientific theory can even be considered so strong that some people will refer to theory as "law", as in the "law of gravity", but this is an exaggeration. Hypothetically, evidence can always be produced that might

4 https://www.youtube.com/watch?v=rA7i9XoeOu4

falsify the theory and cause it to be discarded. Even well-established theories like gravity or germ theory could hypothetically be negated by new and better evidence.

Mathematics is a tool invented by humans that allows us to apply measurements, which, among other things, helps us understand abstract concepts like fine-tuning. Without mathematics, science would be ineffective, and we would be unable to quantify the universe. Mathematics and logic are directly related—in fact, there is a category of mathematics called mathematical logic that studies formal logic within mathematics. Where science has hypotheses and theories, mathematics has theorems and proofs. A mathematical proof is a deductive argument for a mathematical statement, and a theorem is a statement that has been proven. For example, the Pythagorean theorem has over three hundred fifty known proofs. For a valid proof to exist, every statement must have a supporting reason, which in itself becomes another statement requiring a supporting reason, which creates an infinite regress problem and means that proof is actually unattainable. For every valid proof in mathematics, we need to accept that logic exists, and we need one or more axioms, which are statements *accepted without proof*.

Here's where logic plays an integral role. In a lecture, Stephen Hawking once said that hypothetically, you could have been created yesterday by some form of artificial intelligence and implanted with false memories of a longer lifetime, and there is no way to prove that hypothesis is false. Okay, hypothetically it *might* be true, but it is far more likely that the hypothesis is false. Is it more reasonable to believe that my wife and I were created yesterday, both with false memories of thirty-five years of marriage, complete with children and grandchildren, or that we've really been married that long? Which scenario is more complex than the other? Clearly, the artificial world created yesterday is far more complex than the natural world because of all the planning required. Everyone within our sphere of influence would also have to have been created yesterday, with no conflicting false memories that contradict our programming. It is logical to believe that the less complex explanation is the most plausible.

Statistics is an expression of mathematical probability. If a prediction of success is very likely, there is a high probability of success. For example, your favorite football team might be winning by two touchdowns with less than five minutes to play in the game. The probability of success (your team wins) might be greater than ninety (90) percent. Each second that goes by without the other team scoring, the probability will increase toward one hundred percent. By the time there is less than one minute left on the game clock without the other team scoring, the probability of success might be as high as 99.9999 percent—but it can never become one hundred percent until the game is over. Only then can the final result be called a certainty—when there is no doubt in regard to the outcome. There is a difference between 99.9 percent probable and 100 percent certain. Even though the difference is slight, it is quite real.

We don't know... it's simply not possible to know what might have existed before the Big Bang. All we can do is guess. Was there a previous universe? Quantum foam? An infinite number of strings waiting for matter to form? Or was there nothing at all? We don't know, and more importantly, we can't know. We'll never know. But what we do seem to know is that there was a point in time when no stars, galaxies, or planets existed. That seems to be a pretty big deal.

The Big Bang represents the moment in time when the universe got jump-started and cosmic inflation occurred in the blink of a divine eye, either immediately before or immediately after the Big Bang. Both anomalies were executed with incredible mathematical precision. Physicist Sean Carroll says that the fine-tuning argument is a terrible argument because we don't know what might have happened if the cosmic parameters that Martin Rees said were fine-tuned were dramatically different—perhaps a universe capable of supporting a vastly different kind of life might be possible. However, we do know that Rees identified six of these parameters and claimed that even the slightest change in any of the six would cause life as we know it, and perhaps even the universe itself, to fail or not exist. The universe that Carroll speculates *might* exist if the parameters were dramatically changed is hypothetical, but the universe we live in actually exists.

There is always another potential explanation for any phenomenon. People who might object to the true dichotomies that the universe is either planned or unplanned and God exists or God doesn't exist may also have no problem with the claim that Darwin's theory of evolution is the only possible explanation for the evidence of comparative anatomy, DNA analysis, and the fossil record is to conclude evolution is the right answer, and the only answer to the question of how modern life came to be. Evolution is one possible explanation, except in the planned versus unplanned scenario that process is unplanned, and thus not the only possible explanation. It is also possible that animals were created in the manner described by the Bible, and it is possible that biblical creation is wrong but intelligent design is true (intelligent design could be the same processes as those claimed for evolution, but occurring by design or plan rather than being unplanned and undirected). Intelligence is responsible for the existence of life, the universe, and everything, or it isn't. It is possible to frame existential questions so that only binary alternatives are possible.

Each of the seven miracles identified and defined in this book represents an extremely low probability of success, even though all seven miracles must occur in order for you to even exist. Success is essential, but the probability of success is ridiculously low. This is not my determination; it is a derivative of the work of the experts quoted in this book. However low the probability of success for each of these miracles might be, any alternative potential solution is even less likely because that solution must be unplanned and undirected, which means luck must be involved, and by definition, luck dramatically reduces the probability of success. It is especially important to stress that while the probability for the success of each miracle is extraordinarily low, all seven miracles *must* be successful in order for you to exist, and you must exist to read my book and contemplate the problems presented. Indeed, you are literally the final miracle of the seven listed.

My worldview is binary...God exists, or God doesn't exist. The universe is either planned or unplanned, or you could say the universe is either directed or undirected. God made the universe, or some really unbelievable good luck made it. Intelligence was involved, or it wasn't. The interesting problem that arises from contemplating these questions is whether intelligence can arise from non-intelligent origins. One thing is certain; *nothing* did not create the universe. *Nothing* is not a viable explanation for anything. Nothing literally means "no thing." In fact, nothing is the absence of a viable explanation. A force created the universe. That force was either intelligent or not, and that intelligence formed a plan...unless there was no plan.

Any potential third option will still fall under the umbrella of the two basic options: planned or unplanned. A "partial" plan is still a plan. The natural state of an unplanned environment is chaos. Yet evidence of order is ubiquitous. As far as probability is concerned, none of us should exist. And yet, we do. Seven miracles explain our existence. Should any of the seven miracles fail to occur, life as we know it would not exist, and perhaps even the universe would not exist.

And you would not be around to read this book.

THE RELEVANCE OF LUCK

I've been known from time to time to make provocative comments such as, "Time and Luck are the gods of methodological naturalism," mostly because that statement is true, and it annoys my atheist friends when I do. In each of the sections on the seven miracles, there will be a few paragraphs discussing the impact of planned versus unplanned events. To be clear, unplanned events require a range of luck to succeed, but planned events require no luck at all. The question becomes how much luck is acceptable? And how much luck is necessary? Each miracle mentioned in this book has some estimated probability of success assigned by the experts in mathematics who understand the science. Roger Penrose calculated the probability of the fine-tuned universe. Fred Hoyle calculated the probability of an individual cell coming into existence due to random processes as 1 in 10^40,000th power, an absurdly low chance of success. These low chances of success represent the amount of luck needed to achieve just one of the seven required miracles.

In his book *The Blind Watchmaker,* the well-known atheist Richard Dawkins wrote, “We can accept a certain amount of luck in our explanations, but not too much. The question is, how much? The immensity of geological time allows us to postulate more improbable coincidences than a court of law would allow. But even so, there are limits.”[5]

Indeed, there are limits. Dawkins just admitted that Time and Luck are both crucial for his theories to become plausible. The problem is that humans don’t get to say what these limits on luck are. They are dictated by the amount of luck necessary for the unplanned event to succeed. Humans don’t get to draw some arbitrary line in the sand and say only this much luck is acceptable. If you don’t like Hoyle’s calculations, produce new calculations based on more realistic numbers. The problem with that strategy is that we know more about the cell today than we knew when Hoyle was alive, and the odds aren’t getting any better.

---

5 Dawkins, Richard. *The Blind Watchmaker*. WW Norton Company. New York. 1996. Print.

In 1997, mathematician and evolutionary biologist Richard Lewontin wrote an article titled "Billions and Billions of Demons" for the New York Review of Books. In the article, Lewontin admitted that he and his colleagues were advocates for secular atheism thinly disguised as scientists:

> *"Our willingness to accept scientific claims that are against common sense is the key to an understanding of the real struggle between science and the supernatural. We take the side of science* in spite *of the patent absurdity of some of its constructs,* in spite *of its failure to fulfill many of its extravagant promises of health and life,* in spite *of the tolerance of the scientific community of unsubstantiated 'just so' stories, because we have a prior commitment, a commitment to materialism. It is not that the methods and institutions of science somehow compel us to accept a material explanation of the phenomenal world, but on the contrary, that we are forced by our a priori adherence to material causes to create an apparatus of investigation and a set of concepts that produce material explanations, no matter how counterintuitive, no matter how mystifying to the uninitiated. Moreover, that materialism is absolute, for we cannot allow a Divine Foot in the door."*[6]

Read any book about science—I don't care if it's paleontology and the fossil record or about synthetic chemistry and the origin of life. If the author is honest, the word luck will appear in that book, because luck is a required feature of a potential explanation for an unplanned and undirected event. Usually, the author will try to diminish and downplay the role of luck in their scenario, but occasionally, a respected scientist and author like Michael Benton will openly admit the significant role that

---

6 Lewontin, Richard. "Billions and Billions of Demons", New York Review of Books, 1997, Print.

luck played in the survival of Lystrosaurus in the aftermath of the Permian extinction in his excellent book *When Life Nearly Died*. Benton wrote:

> *"I explained that Lystrosaurus was a survivor of a sort, but it was not particularly fast, fearsome, or intelligent. It was really like something of a Triassic pig in appearance, and perhaps even in habits, since it was probably a generalist, without specific adaptations in its diet, living requirements, or mode of locomotion. The point is that good fortune is a characteristic of mass extinctions.* ***The survivors are more lucky than specifically adapted.***"[7]

We are left to consider two choices: either luck has been a significant factor in explaining the universe and the world as they exist today, or luck played no role at all because divine intervention has guided the outcome not once or twice, but at least seven times. Once again, it's the unplanned versus planned universe argument. We can either accept that luck has played an enormous role, or luck has never been involved and the universe exists by design.

PREBUTTALS

Before publication, the original, raw, unedited manuscript was provided to a number of reviewers for the purpose of generating constructive criticism. In some excellent feedback, several potential weaknesses in the book were helpfully pointed out. This section is intended to address some of these issues. My manuscript appraisers had several specific concerns. One reviewer pointed out that I said that scientists have claimed to be able to explain everything if we conceded the first miracle, the origin of the universe itself. However, I was paraphrasing

---

[7] Benton, Michael. *When Life Nearly Died*. Page 22. London. Thames and Hudson, Ltd. 2003. Print.

the quote from Joe Rogan about *what scientists have claimed.* Also, I'm pretty sure he didn't mean the claim to be taken literally...it was told as if the claim itself was a joke. Not to be taken seriously. That reviewer also said that concepts like baryogenesis and quantum gravity are not well understood, which is very true. But baryogenesis happened in the early universe, after the Big Bang had occurred. As for quantum gravity, mathematical formulas might suggest it exists, but, like God, no one has ever seen it. Quantum gravity might explain the mechanism that started the universe, but it doesn't answer the *why?* question.

Here are a few other issues raised by my manuscript appraisers:

Argument From Incredulity

This critique means parts of my argument could be rejected because it commits the logical fallacy of claiming something is false because I personally cannot imagine or understand how the claim could be true, substituting my lack of comprehension for actual evidence. I have addressed this problem by adding language that clarifies the difference between a true impossibility and a virtual impossibility.

Specifically, the criticism was this: "Throughout the manuscript, certain scientific events (like abiogenesis or cosmic inflation) are characterized as 'impossible' simply because they are extremely unlikely or not yet fully understood. This can unintentionally shift from scientific critique into 'I can't imagine how this could happen, therefore it didn't.' That style of reasoning might make skeptical readers tune out."

That's a fair point. On a probability scale, there are two extremes. Both of these extremes represent absolute certainty. Let's say, for example, the scale represents belief in God. At one extreme, we have a zero percent probability that God does not exist—which is claiming real knowledge with absolute certainty. At the opposite extreme, there is one hundred percent certainty that God exists, with the same caveat. Can anyone claim to be one hundred percent sure beyond all doubt and have enough confidence to claim one of these extremes? Let's take my own case as an example. Can I be one hundred percent sure that God exists? I say no, despite my years of research and one intense personal experience. The best I can do is approximately 99.999999 percent, give or take a few decimal points. I must allow for some slight chance that I'm bat-poop insane and delusional (as Richard Dawkins suggested), and that a lifetime of false memories might have been implanted in my head. My own experience meeting Jesus Christ one night in my bedroom could have been a pleasant delusion, and nothing more. The paranormal experiences I had as a teenager could have been figments of my imagination or a form of mental illness. The reasons I believe what I believe could all conceivably be false. The question is, do I sound like a mentally ill person? Or is my prose lucid and easy to understand?

As I've read and learned about existential science over the past fifteen years or so, I've developed what I call my "big picture" argument: the universe must either be planned or unplanned. Or directed versus undirected, if you prefer. God exists, or God doesn't exist. It's that simple. In my opinion, we should all act as if we were scientists and treat everything as if it were a scientific theory. It's okay to look at all the evidence (such as the evidence in this book and also in my book *The God Conclusion*) and interpret that evidence to make a decision either to believe or not believe God exists or the Big Bang theory, but we must keep our minds open and receptive to new evidence that could contradict our current worldview. My personal belief that God exists is quite strong, and it would take quite the comprehensive effort to convince me that my current worldview is false because it is based on both my interpretation of scientific evidence and a number of powerful and remarkable personal experiences. For example, I believe ghosts exist, or at a minimum, a form of intelligence exists not visible to the human eye. I believe this because one of my friends during our youth lived in an (allegedly) haunted house that was actually somewhat famous for being haunted. I didn't have one single experience in this house that could potentially have a logical explanation: I had dozens of experiences over a period of several years of such variety that I cannot think of a rational explanation for them unless my friend turned out to be the brains behind the illusionists Penn and Teller.

My problem is that I can describe my experiences to you, but I cannot experience them with you. There is no way, other than you visiting the same house and having the same or similar experiences, that I should be able to convince you that ghosts exist. I can tell you my ghost stories. I can share the ghost stories of other people with you. I could even show you videos that allegedly have captured a ghost on camera. You can simply respond that the stories/anecdotes are all fictitious, that the camera was showing special effects rather than actual images, and dismiss all of my claims without bothering to consider them. You have free will. You don't have to believe me. Believe whatever you wish.

On the other hand, Karl Jung once said that you can claim to have never had experiences like mine, but you cannot claim you know with certainty my experiences were just a figment of my imagination.

Don't simply take my word for anything. Use the hyperlinks in the index to verify that the people I've quoted actually said what I claimed they said. Use that miracle of intelligence, yet another gift from our Creator, and let logic help you decide what is true.

False Dichotomy

A false dichotomy (sometimes referred to as the excluded middle fallacy) is a logical fallacy created when only two options are presented, but other options exist that are disregarded. For example, we might claim that only two colors exist, black and white, while ignoring the fact that a third color called "gray" can be created by mixing black with white. Again, the complaint specifically was this: “Some sections present the choice as ‘either God intentionally designed this’ or ‘blind, unguided chance.’ But many readers know there are intermediate positions–like theistic evolution, guided evolution, non-random selection mechanisms, or natural laws that bias certain outcomes. Treating the range of views as a strict two-option choice may feel too reductionistic.” Another fair point.

My response is that a binary choice or dichotomy only becomes a logical fallacy when a third option can be identified. In a computer, there are only two options that can be stored as individual bits in a byte of information: either a zero or a one. You literally cannot put any other character or number in a bit and have the computer understand or process that information. The computer knows only zeros and ones. Those zeros and ones can be ordered in specific sequences and used to convey information by doing things such as turning the letter “S” or the percent symbol into a sequence of zeros and ones. An individual bit can only contain a zero or a one, never a two. Computer machine language is a true dichotomy, not a false one.

When I say the universe is either planned or unplanned, it was my intention to argue the claim presents a *true* dichotomy, not a false one. I’m not saying the accusation that my book creates a false dichotomy is wrong, but I am saying IF my argument is wrong and there is a third option besides a planned or unplanned universe, my critics should be able to say what that third option might be. Until that time, I’m going to continue working on the premise that my argument creates a true dichotomy as it currently exists, without modification, not a false one. In the big picture, the universe is either planned or unplanned. The most widely accepted definition of the word evolution describes an unplanned and undirected process that begins with the very first form of life and continues through every form of modern life, driven by a series of mutations and adaptations that allow for incredible metamorphosis if enough time can be claimed to have elapsed. Theistic evolution and guided evolution are synonymous terms that both represent an oxymoron—theistic evolution is claiming to be a planned nonplan. "Non-random selection mechanisms" is evolution-speak for mating habits. Using the term "theistic evolution" is a compromise I'm not willing to make because I would have to abandon my big-picture argument without reason for doing so other than to make others happy.

Theism is claiming that there is a plan. Evolution is an argument that no one is directing nature. It's like trying to mix oil and water. Now, here's the tricky part—you can believe and claim that processes attributed to evolution theory such as non-random selection mechanisms and provide powerful evidence these effects exist, but if God is being given the credit, you can't call the process "evolution" if God is involved, even if you believe "monkeys make men", as Darwin so famously scribbled in his notebook. Call it what it is: intelligent design.

There are not two bodies of evidence. There is one body of evidence and two interpretations. Planned or unplanned. Directed or undirected. God or no God.

Overgeneralizations About Atheists and Scientists

The advice offered suggests, "Phrases suggesting evolution is a 'religion,' or that critics want to 'burn heretics,' can come across as ad hominem or exaggeration. These statements distract from the strength of your arguments and may close off otherwise curious readers."

Once again, great advice. My only defense is that we are talking about the first rough draft. I intend to do some wordsmithing to dull the rhetoric and soften the tone a bit. I'm about to go back through the book from cover to cover, looking for ways to tone down my rhetoric to not sound so bombastic. I decided to revisit the marked phrases to see if and how I might edit them to make the book more palatable to the average reader. I tend to get rather opinionated when sharing my thoughts about some of our more outspoken atheists, such as Sam Harris and Richard

Dawkins, and I do need to make a concerted effort to either tone down my rhetoric or make sure my thoughts are crystal clear. I've gone through the manuscript and edited passages that sounded unnecessarily provocative.

Heavy Reliance on Analogy Over Evidence

Another perspective offered suggests, "Analogies like watches in the woods or binary code in DNA make intuitive points, but they're not logically equivalent to the systems they represent. Sometimes the argument leans too heavily on these metaphors rather than on data about biological processes. Analogies are good - and even important - as illustrations to clarify a concept, but they can become misleading if they're treated as direct evidence rather than as tools to help the reader understand the underlying idea."

Wow. If all my feedback is this good and I follow such sage advice, I'm going to end up with a great book.

I am inclined to agree; analogies are typically imperfect and difficult comparisons to make. For example, Paley's famous Watchmaker analogy was (allegedly) refuted by the claim that complexity doesn't require a designer. According to people like Richard Dawkins, evolution can explain complexity in nature without invoking an intelligent Creator. However, my book does present evidence, not merely anecdotes. The evidence is largely presented in ordinary English rather than technical jargon because, in my opinion, communication is essential. I want to write a book that everyone can read and understand, not one that causes the eyes to glaze over for long stretches.

The word evolution simply means "change", which means the change of an existing thing. Creation means the introduction of a new thing that didn't exist before, and life cannot evolve until it exists. Before evolution can even become possible, creation has already occurred. The only question is whether we give God or nothing the credit. I have two decades of experience as a software developer. I know computer software fairly well, though my skills have atrophied somewhat since I stopped writing code for a living. Nevertheless, I still remember quite a bit, such as how the computer internally sees data and processes that data to become information. I'm not an expert in genetics. However, as the commercial says, "But I did stay at a Holiday Inn Express last night."

By that I mean I've read a book or two on genetics and know a little more about DNA than nothing. For example, I know that DNA is considerably more complex and restrictive than the binary machine language stored in the computer. I know that books can be digitized, stored, and retrieved from DNA. I know that DNA from a human being is more complex than a computer's operating system. I certainly don't know everything, and arguably much of anything, but I do know a few things, and I'm happy to share bits of information I've learned from reading books written by smart people. As I read through the manuscript in additional editing passes, I promise to be on the lookout for passages that rely too heavily on anecdotes or analogies and hope to be able to substitute more evidence instead.

I prefer to quote actual experts rather than try to pass myself off as one. I merely take the ideas of other people and apply logical thought to their claims to form my opinions, mostly by standing on the shoulders of modern-day giants.

MIRACLE 1
THE ORIGIN OF THE UNIVERSE

INTRODUCTION

Not that long ago, there were two competing ideas about the origin of the universe. The first idea, called the steady-state (or eternal universe) theory, claimed there was no beginning; the universe was thought to be eternal and had basically existed forever. The second theory was that the universe had a beginning, an idea that probably originated from a theological perspective. For a time, there wasn't sufficient evidence to favor one side of the debate over the other. Even so, the advocates of the steady-state theory felt confident enough in the logic of their beliefs that they mocked and ridiculed the alternative, but the evidence didn't support their beliefs any better than Big Bang cosmology.

Astronomer Fred Hoyle actually named the other hypothesis "the Big Bang" as a joke to show his utter contempt for the idea. However, virtually all available evidence supports the Big Bang, and the hypothesis eventually became theory. Perhaps the resistance to the Big Bang stemmed from the idea that the universe has a beginning, creating an essentially unsolvable problem for a purely naturalistic worldview. Scientists now try to differentiate between the Big Bang and the origin of the universe, saying that the singularity occurred when the universe was in a hot, dense state and no stars, planets, or galaxies existed. The true origin of the universe may have occurred immediately before the Big Bang, or at some other time. Some have even argued that the multiverse could be eternal, even though our universe has not always existed. Scientists can only look so far back into the past. They cannot look back in time to see what might have existed before the Big Bang.

“Nothing” is one option, albeit a somewhat incomprehensible one. A previous universe that fully expanded and then collapsed back into nothing (a cyclic or bouncing universe) has been proposed as another option. A third option is an eternal multiverse in which our universe popped into existence from a static parent that has always existed. No material in our universe today existed before the Big Bang, as far as we know (or believe we know), but the multiverse might hold the building blocks of our universe. But we can never observe any evidence for the static parent universe because to do so, we would need to be able to look outside of our universe. What do these multiple-choice options have in common? At the instance of the Big Bang, this universe contained no things—no stars, planets, etc. Not even one atom. The evidence that appears to show this is considered to be quite strong. In all three scenarios preceding the Big Bang, nothing in our current universe existed. Whatever might have existed, it wasn’t matter. It was infinitely denser than matter.

Among contemporary physicists and cosmologists, there isn’t much serious argument today about whether the Big Bang actually occurred because the evidence for the Big Bang is considered to be quite strong. Instead, the physicists and cosmologists argue about what might have potentially existed prior to the Big Bang. Scientific evidence confirms the Big Bang. Question: How do we get a universe full of matter and a planet teeming with living organisms from literally nothing? What started the expansion of this material that apparently came from nowhere? Also, what started and stopped cosmic inflation? What created life from inanimate matter? The transition from non-life to life was a pretty big one. And what made life intelligent?

It has been widely accepted that the evidence shows that if we look backward in time, the universe keeps getting smaller until it condenses to a hot, dense state where no stars, galaxies, or planets existed (literal nothing). All the matter of the universe was condensed to a point where it was smaller than the tip of a pen, and from that point (which scientists call t=0) the universe began to expand. As we reverse direction and look forward in time with t=0 as our starting point, the universe began to rapidly expand and gradually become the place where we live today. Scientists claim to know (and thus should be able to explain) practically everything that happened in the universe immediately following the Big Bang from mathematical calculations, but their equations can't explain the Big Bang itself. Some physicists and cosmologists today like to speculate about the origin of the universe and say that the Big Bang was not the true origin. They believe *something* must have existed prior to the universe, either a prior universe that expanded and then contracted back to nothing, or a multiverse. Admittedly, this approach of denying the Big Bang as the origin of the universe would appear to solve one of the big problems caused by the Big Bang—how did our universe come from nothing? The simple answer: it didn't. It came from God.

It's important to note that just because an answer might be easy, it doesn't mean that answer is wrong. Some might argue that I've simply inserted an argument for the God of the gaps, but that would only be true if I merely threw up my hands and said, "God did it!" without giving the problem any serious contemplation. These are not mere gaps; they are information black holes. There will never be a material explanation for the origin of matter. Humans will never understand how life was created from inanimate matter. Humans will never understand why they are conscious about the fact that they don't even understand the questions well enough to properly answer them. Even so, that doesn't mean we can't figure out a few things.

But we're never going to understand everything.

Understanding Nothing

Before going any further, we should first explore what the word nothing really means: literally no thing, which means not even atoms or “empty” space filled with oxygen molecules. Nothing is admittedly a tough concept to grasp. When we often claim to see nothing, there is still *something* there. It might be air molecules, or it could be only hydrogen atoms, but something is definitely there. You can’t see *nothing* with the naked eye. It’s difficult to even conceive of true nothingness.

Imagine you live in a city and wish to drive to your friend’s house in the suburbs. You program his street address into your GPS and follow the directions, but when the GPS announces you have arrived at your destination, you’re sitting in front of a vacant lot. You might think, “There was nothing at that address,” but is that true? There might have been grass and trees in the vacant lot, or it could have been an asphalt parking lot, and there certainly was sky and dirt. Literal nothing literally means no birds, no bees, no worms, no trees, no people, no grass, no weeds, no earth, no air, and no water. No *things*, not even a vacuum. Matter is something, not nothing. Space is something, but nothing is literally nothing. The concept is virtually impossible to grasp. Even if we hold our thumb and forefinger an inch apart and claim there is nothing between them, something is there. Air molecules are there. Space is there. *Something* is there, even if you can’t see air molecules.

It is impossible to see nothing, because you’re always looking at *something*. Even if you don’t think you see anything, atoms and molecules are floating around, invisible to the naked eye. Yet before the Big Bang, neither time, space, nor matter allegedly existed. Nothing existed. Nothing currently in this universe. We could argue for eternity about what might have existed prior to the Big Bang, but the only thing we know for certain is what *didn’t* exist—our universe in its current form. By playing semantic word games, we can entertain endless possibilities with baseless speculations. We can believe an infinite number of "universes" in a multiverse hypothesis were created, but failed immediately to compensate for the sheer improbability that this allegedly fine-tuned universe came to exist purely by chance and due to unplanned and undirected processes. We can choose to believe that the material for our universe has always existed and expands and contracts, kind of like an accordion, over very long periods of time. We can make up any crazy and unprovable theory we'd like to explain what existed prior to our universe and why our universe exists. But what we cannot do is deny the evidence for the Big Bang itself, or the fact that prior to the Big Bang, no stars, planets, galaxies, etc., currently part of *this* universe existed.

Some people like to criticize Kent Hovind because he was convicted of tax fraud for not paying federal income taxes after declaring his work constituted a Christian ministry, but the IRS disagreed. Hovind did lose his court case and spent several years in prison, effectively paying his debt to society. While Hovind may not have a perfect understanding of the U.S. tax code, his understanding of existential science is outstanding. Kent Hovind is not perfect. Nobody else is either. All have sinned and fall short of the glory of God. Regardless of his legal issues, Hovind squared off in a live debate against three atheists simultaneously and virtually decimated their arguments. He said this about the origin of the universe:

> *"Time, space, and matter are what we call a continuum. All of them have to come into existence at the same instant because if there were matter but no space, where would you put it? If there were matter and space but no time, when would you put it? You cannot have time, space, or matter independently. They have to come into existence simultaneously. The Bible answers that in ten words: In the beginning—there's time—God created the heavens—there's space—and the Earth. There's matter. So, time, space, and matter created a trinity of trinities. Time is past, present, and future. Space has length, width, and*

height. Matter has solid, liquid, and gas. You have a trinity of trinities created instantaneously, and the God who created them has to be outside of them. If He's limited by time, He's not God. The guy who created this computer is not inside the computer. He's not running around in there changing the numbers on the screen. The God who created this universe is outside of the universe. He's above it, beyond it, in it, through it. He's unaffected by it. The concept that a spiritual force cannot have an effect on a material body—you'd have to explain to me things like emotions and love and hatred and envy and jealousy and rationality. If your brain is just a random collection of chemicals that form by chance over billions of years, how on earth can you trust your own reasoning processes and the thoughts that you think?"[8]

Whether you love Hovind or hate him, you ought to concede that was a brilliant answer. Nevertheless, people will reject the argument simply because of the source. Okay, then how about Neil deGrasse Tyson? Would you listen to what he says and believe him? Tyson said:

"You ever wonder what nothing is? Is this nothing? No, it's air. Let's go where there's no air. How about outside our atmosphere? No, there's still a few particles floating there between the planets, so it's not quite nothing. Between the galaxies there is even less, but still a few particles per cubic meter that lurk there. Alright, let's imagine you can get rid of those. Now what's left? In the pure vacuum of space, there's something called virtual particles that pop in and out of existence. Quantum physics tells us this. So, there's still something there. To get a true nothing, you have

---

[8] Hovind, Kent. "Kent Hovind – Space Time Matter", Smiley, December 13, 2018, https://www.youtube.com/watch?v=cTFuRMTyvCQ

> *to go where there's not even space or time. But if the laws of physics still apply, then there's still something there. So, you have to go where there's not only no matter, and no space/time, but no laws at all. Behold, a true nothing."*[9]

Basically, what I just said.

Unnecessary melodrama and questionable acting aside, in this instance, Tyson does a decent job of communicating the problem so that it is more easily understood, although I would quibble that the "laws of physics" don't apply if physical matter doesn't exist. Tyson loses me when he's asked about "the scariest fact" he knows, and Tyson's answer is that in 22 billion years, the universe may have accelerated to the point there is a chance that atoms will come apart and the universe ends abruptly in what he calls "The Big Rip." After another 22 billion years have passed.

For proper context, the universe is currently less than 15 billion years old according to the most generous estimates, which means it is less than halfway to this hypothetical universe death. Assuming that the hypothetical end of the universe does come to fruition, by that time, we will have all been dead for approximately 22 billion years. We're talking about a really, really long time from now. The universe hasn't even experienced a midlife crisis, and Mr. Doom-and-Gloom is worried about an event that won't happen *until he's been dead for 22 billion years*. Not exactly a scary proposition, is it?

Conversely, when entertaining the same question, my personal worst fear is a bit more pragmatic. My worst fear is that nuclear war could break out between two somewhat evenly matched

---

9 Tyson, Neil deGrasse. "Nothing is Something w/Neil deGrasse Tyson", Universal Law, May 3, 2025, https://www.youtube.com/shorts/0vgR5qN0xmM

nuclear powers, for example, say India and Pakistan. The two countries consider each other enemies, share a common border, and both have nuclear weapons. My fear is that one of them might launch a nuclear strike on the other and initiate a global nuclear holocaust. Or the current conflict between the U.S. and Russia over Ukraine could develop into a nuclear war. Or China could attack Taiwan and start a war. Or Iran could attack Israel. Or a massive volcanic eruption could disrupt the global climate for an extended period and precipitate a mass extinction event. Or the global magnetic poles could suddenly reverse polarity, wreaking havoc. Or a meteor could strike Earth, causing mass extinction. Or Jesus could return to Earth, and nobody looks busy. Something that *might* theoretically happen in another 22 billion years isn't nearly as scary as something that could theoretically happen tomorrow.

There are hydrogen atoms and oxygen molecules you can't see or feel, and yet they exist. It's critically important to remember the word *nothing* literally means "no thing" as in no plants, animals, objects, not even one single atom. The word vacuum is often used synonymously with nothing or to represent nothing, but a vacuum is still *something*—it is space without matter. As I said, nothing is a surprisingly difficult concept to grasp. Space, time, and matter did not exist prior to the Big Bang, assuming the experts are correct. With contrived speculation, we might solve one or two of the problems stemming from the Big Bang, but not all of them. For example, the multiverse could solve the fine-tuning problem by introducing an infinite number of failed or otherwise unobservable universes to offset the peculiar, fine-tuned characteristics of this universe we occupy, but it doesn't explain why the Big Bang happened in the first place. Multiverse hypotheses primarily exist to mitigate the improbability of our fine-tuned universe—if every other conceivable universe also came into existence but failed to persist, then the fine-tuned aspects of this universe become less special. The cyclic universe appears to address the origin of matter problem by essentially making matter eternal, but the First Cause question remains unanswered. It doesn't really solve the problem...it merely attempts to solve one aspect of the problem, in theory.

Science communicator Arvin Ash produces informative videos that explain complex scientific concepts in easy-to-understand language. He's a terrific source of information. About the Big Bang theory, Ash said:

> *"The Big Bang wasn't an explosion in space like when a bomb detonates. It was an explosion of space itself. And it's not clear there was anything beyond this expansion."*[10] *What, if anything, existed prior to the Big Bang? Arvin Ash just conceded it is possible that nothing existed—no multiverse, no previous cycles of a universe that expanded and collapsed. There was nothing, and then there was something. Why? What caused the Big Bang? Ash explains probability, saying, "Essentially, if you roll the dice enough times for nearly eternity, you'll eventually hit the jackpot and a universe will form. In this case, what came before the Big Bang would not really be nothing because quantum mechanics would be that something. In other words, the laws of quantum physics have to exist in order for this to happen."*[11]

What does "the laws of quantum physics" even mean? Do so-called laws of physics apply when matter doesn't exist? Furthermore, from where do these laws come? Did the physical laws which govern our universe exist prior to the Big Bang, when nothing physical existed? Approximately 13.7 billion

---

[10] Ash, Arvin. "What was there before the Big Bang? Three Good Hypotheses!", Arvin Ash, April 13, 2024, https://www.youtube.com/watch?v=QAuf97BhgRY

[11] Ash, Arvin. "What was there before the Big Bang? Three Good Hypotheses!", Arvin Ash, April 13, 2024, https://www.youtube.com/watch?v=QAuf97BhgRY

years ago, the entire universe was condensed into a hot, dense state that was allegedly smaller than a single atom, and for some unknown reason, that condensed material began to rapidly expand and cool off. While there can be much argument over what might have existed prior to the Big Bang and whether time, space, and matter were all created at once, the moment of the Big Bang itself is not up for dispute. It can be debated whether this universe is part of a much larger multiverse or simply popped into existence from non-existence due to quantum mechanics, but there isn't much in the way of serious dissent to the idea that this universe has not always existed in its current form, but mysteriously formed billions of years ago, with remarkable precision and accuracy.

The observed evidence doesn't seem to support a cyclic universe because the universe is still expanding, and after 13.7 billion years of expansion, it probably should have started slowing down by now. But it hasn't. The multiverse is an attempt to solve the sheer improbability problem that this (allegedly) fine-tuned universe we occupy could have come into existence in utter defiance of the laws of probability, and gives us a pseudo-explanation for the origin of the material that expanded during the Big Bang, but it doesn't give us a reason for the Big Bang itself—what caused it? Quantum mechanics suggest that a virtual particle can pop into existence, but an entire virtual universe popping into existence all at once? No matter how you slice it, the universe is not a hypothetical construction. We wouldn't exist if our universe didn't exist. It matters, pun intended.

Arvin Ash also said,

> *"Basically, what the [Big Bang] theory says is that the universe today is bigger than it was yesterday, and much bigger than it was a billion years ago. So, as you extrapolate back, the universe gets smaller and smaller, denser and denser, and hotter and hotter. So, as you keep going, you get to a very small volume of*

> *space which is very dense and very hot. At some point in this extrapolation, our equations stop working because the volume becomes zero."*[12]

When mathematics fails, imaginations run rampant. Admittedly, it is extraordinarily difficult to properly conceive of true nothing.

Neil deGrasse Tyson has speculated that the improbability problems associated with our universe could be resolved if we lived in a computer simulation like the movie *The Matrix*, which happens to be Tyson's favorite science fiction film. Tyson suggested that our life experiences could all be false memories, and he seemed to be quite serious when he said it. In *The Matrix*, Neo was literally kept in a pod, exactly like a brain in a vat, before taking the red pill. Does it make more sense to believe we are living a "brain in a vat" experience in which everything that happens in our lives is simply a figment of our imagination, or that our experiences are real? Even if you suspected the brain in a vat scenario might be true, you can't live your life as if that is reality. Also, how does an even more complex explanation work better than the one we have? How does a giant simulation hiding a complex but dystopian reality work better than simple reality without a synthetic atmosphere filled with random numbers on a hidden green screen flying around and filling the air? If we really lived in the matrix, why don't I see numbers flashing across the screen? Why can't I dodge bullets like Keanu Reeves?

Because it is make believe. Keanu Reeves isn't bending over backwards and dodging real bullets. It's all acting and special effects. Now that we know the universe hasn't always existed, we can ask the question whether the Big Bang and cosmic inflation were planned or unplanned. A computer simulation

---

[12] Ash, Arvin. "Cosmic Inflation; The Solution to the Big Bang Theory and the Universe", Arvin Ash, February 26, 2022, https://www.youtube.com/watch?v=q5dy-NtVeF0&t=202s

might not represent a theological interpretation of the origin of the universe, but it is a *planned* universe. Frankly, an unplanned universe isn't logical and doesn't match the evidence very well, but neither does the matrix, which is merely another version of a planned universe. Acknowledging the potential existence of the matrix begs the question: who created it? Obviously, an intelligent designer with a superior intellect. Why was the matrix created? That answer also turns out to be easy to discern—to entertain us. But what is the point of the matrix? Unknown. To confuse us, perhaps? The biggest problem with the matrix is that whoever is decreed the creator of the matrix must exist outside the matrix. This makes the architect of the matrix indistinguishable from God in terms of the planned universe argument.

EVIDENCE FOR THE ORIGIN OF THE UNIVERSE

The idea is that if you could go back far enough in time, the universe would continue to get smaller and smaller until whatever material that existed immediately prior to the Big Bang was infinitely compressed and flattened to the point where the material that would become the building blocks of our universe could fit into an incomprehensibly small space, akin to the head of a pin. Although there is disagreement about the details, most physicists agree with that idea.

In his book *The Origin of the Universe,* astronomy professor John D. Barrow wrote:

> *"Despite the confidence with which some modern cosmologists have addressed questions about the origin of the universe—a confidence that has seen the publication of research papers bearing titles like 'The Creation of the Universe out of Nothing'—one should be cautious. All these theories need to assume at the outset the existence of a good deal more than one's everyday conception of nothing in order to say anything of interest. In the beginning, there must have*

> *been some laws of nature, energy, mass, and geometry. And, of course, underpinning everything seems to be the ubiquitous world of mathematics and logic. There needs to be a considerable substructure of rationality before any complete explanation for the universe can be erected and sustained. It is this underlying rationality that most theologians emphasize when questioned about the role of God in the universe."*[13]

Dr. Barrow's admonition is worth considering. Why do we assume mathematics and logic have always existed? They are ideas born in the minds of people. Can you take a walk on the beach and stumble across a flawed syllogism? Could you walk into a market and buy the number five? If no humans existed, would the number five even exist? Can you describe a five? If we're talking about currency, of course, you can. Abraham Lincoln's face is on the front of the bill, but the bill is not the same thing as the number it represents. The number itself is only a concept. It's an idea. Literally, the number 5 isn't the concept itself; it's a symbol that represents the idea.

Whatever the material that expanded at the precise moment of the Big Bang was, that material was not a star, planet, galaxy, or atom because prior to the Big Bang, and we can be reasonably sure that not even one single atom existed. We can be reasonably confident of this belief because the vast majority of experts of note on the Big Bang theory has consistently described this same condition of the early universe. Some experts have speculated that this material was the residual debris of a previous universe that collapsed and claim the universe repeats cycles of expansion and contraction, but this speculation cannot be confirmed with evidence, just as the multiverse cannot be confirmed. The Big Bang *is* supported by observable scientific evidence. Multiverses and cyclic universes

---

[13] Barrow, John. *The Origin of the Universe*. Page 110. New York. BasicBooks. 1994. Print.

are only supported by math equations and wishful thinking. They are hypothetical solutions to very real problems.

Barrow adds:

> *"That part of the scientific enterprise that tries to explain the existence of the universe as a consequence of a prior state consisting of* absolutely nothing *violates our deep-rooted sense that 'there is no such thing as a free lunch.' Nonscientists take it for granted that you can't manufacture something out of nothing. If one proposes to give a scientific account of a universe coming into being, an immediate objection seems to be that one would indeed be trying to get something out of nothing, because one would have to bring into being a universe that possessed energy, angular momentum, and electric charge. This would violate the laws of nature, which enshrine the conservation of these quantities, and so the creation of the universe cannot be a consequence of those laws."*[14]

We can either take the universe for granted and assume it had no choice but to create itself, or be grateful for the miracle as the first of seven total gifts from God. Statistically speaking, the universe should not exist. In truth, the Big Bang sounds a lot like a spontaneous generation theory, not just for life but for the universe itself. The origin of the universe occurred. Scientific evidence tells us this. The only remaining question is what might have existed prior to the Big Bang.

Whatever might have existed, we know that it wasn't this universe.

---

[14] Barrow, John. *The Origin of the Universe*. Page 111. New York. BasicBooks. 1994. Print.

Redshift

Astronomers had discovered that light would appear to be different colors depending on whether a celestial object was moving closer to or further away from our point of observation. If an object was getting closer, the light would appear bluish, and if it was moving further away, the light would appear redder, thus the term *redshift* was coined to describe objects in the sky moving further apart. This was the first piece of scientific evidence that validated the Big Bang theory.

Austrian mathematician Christian Doppler first hypothesized an explanation for the phenomenon all the way back in 1842, predicting the colors of stars could be due to their movement relative to Earth. Then in 1912, Astronomer Vesto Slipher observed blueshift in relation to the Andromeda galaxy, indicating it was moving closer to Earth. Next, in 1923, Edwin Hubble developed Hubble's law to estimate the speed of any galaxy moving away from Earth. Again, these were the very first pieces of evidence confirming the Big Bang theory. Hubble's observation became the first critical piece of information confirming the universe had an origin. Cosmic Microwave Background Radiation was the final piece of scientific evidence that confirmed the event known as the Big Bang, and the hypothesis became a scientific theory.

Cosmic Microwave Background Radiation (CMB)

In 1964 physicist Arno Penzias and astronomer Robert Wilson discovered the cosmic microwave background radiation claimed to be residual evidence of radiation or leftover heat originating from the Big Bang. Together with redshift, CMB is another piece of the evidence that cemented the move for the Big Bang from a hypothesis to a widely accepted scientific theory. Cosmic Microwave Background Radiation is the name given to the faint background glow representing the leftover heat created by the combination of the Big Bang singularity and cosmic inflation. The CMB has a uniform temperature that fluctuates slightly across the entire universe.

In the decades following the discovery by Wilson and Penzias, fresh evidence continues to be discovered and observed that further confirms the two theories. This new evidence has only strengthened the theories.

Fine-Tuning and Multiverse

The Cosmic Background Explorer (COBE) satellite, also known as Explorer 66, operated from 1989 to 1993. COBE collected data about the CMB emitted by the Big Bang. According to the Nobel Prize committee, George Smoot and John Mather were awarded the 2006 Nobel Prize in Physics for "discovering the blackbody form (a physical body that absorbs electromagnetic radiation) and anisotropy (directionally dependent) of the cosmic microwave background radiation." I'm not going to pretend I can translate that sentence beyond what I've already provided. I don't know what a blackbody form is or what it means for the CMB to be directionally dependent. The gist is that their work provided additional confirmation of the Big Bang because observations from the satellite matched predictions of the theory. Additional data about the cosmic microwave background have been collected more recently by the Wilkinson Microwave Anisotropy Probe (WMAP) and Planck Space Observatory, providing further confirmation of the Big Bang theory. Interestingly, images from the James Webb Space Telescope seemed to indicate massive galaxies were present in the very early universe. These images appeared to conflict with the Big Bang, but a more detailed examination of the data led to the conclusion the images did not conflict with the theory after all.

I have also read of claims saying that measurements of deuterium abundance in the early universe provide even more support for the Big Bang theory, whatever that means. Redshift and CMB are sufficient evidence to move the Big Bang from hypothesis to theory without the introduction of additional supporting evidence, but more evidence never hurts. Together with the evidence for redshift plus all the research into the cosmic microwave background from COBE, WMAP, and the Planck observatory, the belief that this universe originated from a hot, dense state at a time when no stars, galaxies, or planets existed seems to be firmly and overwhelmingly established by scientific evidence. There isn't a reason for extended debate until some new earth-shaking evidence causes us to reverse course and believe the universe itself could be eternal. Without this additional evidence, there should not be any serious dispute over the claim that there was once a time when no stars, planets, galaxies, or atoms existed. All the scientific evidence clearly shows that the universe and life have not always existed. Our universe had a beginning. It seems rather silly to acknowledge that atoms have not always existed but then to suggest the universe might have somehow existed before space, time, and matter could be created. It appears to be a semantic hypothetical argument to suggest the universe might have existed prior to space, time, or matter existing—a problem that only exists because of a universe created from absolutely nothing. A universe perfectly suited for producing abundant life on at least one planet.

Evidence of Fine-Tuning

The fine-tuning argument makes the claim that precisely defined cosmological variables exist and if any of them were even slightly altered, life and potentially the universe would fail to exist. Lord Martin Rees wrote a book titled *Just Six Numbers,* which identified six cosmological factors on which the success or failure of the universe depends.

These are the (allegedly) six cosmological factors that the slightest change to any of them would cause the universe to fail. In his book *Just Six Numbers* Lord Rees claimed six very precise cosmological factors determined the success of the universe. These factors are:

1. Omega (value = 1) to represent the amount of matter in the universe.

2. Epsilon (value = 0.007) represents the degree to which atomic nuclei are bound together.

3. "D" is the number of dimensions (value = 3)

4. "N" represents the strength of electrical forces that bind atoms, divided by the force of gravity (value = 1,000,000,000,000,000,000,000,000,000,000,000,000,000)

5. "Q" represents two fine-tuned fundamental energies (value = 1/100,000)

6. Lamba (value = 0.7) represents anti-gravity in the universe

Roger Penrose calculated the probability of the fine-tuned universe created by the Big Bang and cosmic inflation as an absurdly improbable number that represents an extreme improbability which might be called by some a virtual impossibility, something on the order of 1 in 10 in 10^{124}. This is an extraordinarily low chance of success. Let me put it this way—no one in their right mind would ever visit Las Vegas and place a bet with those odds going against you. A one percent chance of winning or success is significantly higher by many orders of magnitude.

Astronomer Fred Hoyle famously said, "A common sense interpretation of the facts suggests that a super-intellect has monkeyed with physics, as well as with chemistry and biology, and that there are no blind forces worth speaking about in nature. The numbers one calculates from the facts seem to me so overwhelming as to put this conclusion almost beyond question."

In his book *The Living Cosmos,* physics professor and writer Chris Impey wrote:

> *"Suppose a deck of cards represented randomly selected atoms in the universe. In one deck of cards, the aces would be helium atoms, and the other 48 would be hydrogen atoms. You'd need 30 decks of cards before you'd expect to find one carbon atom. You'd need to search 300 decks to find a single iron atom. How do we know what the universe is made of? Astronomers use remote sensing, specifically spectroscopy, to measure the composition of star stuff. Each element has a unique set of sharp spectral features that acts like a fingerprint, so by identifying that fingerprint in starlight, astronomers can measure contributions of distinct elements."*[15]

The interesting thing is that all these people performing the calculations happen to be atheists, yet they really don't like the potential odds of success for an unplanned universe. Instead, they concoct these strange ideas like multiverses and cyclic universes to resolve troublesome problems posed by a universe created from nothing and life arising from inanimate matter.

Physicist Sean Carroll also happens to be an atheist, and he has some very interesting ideas about the Big Bang. Carroll has said,

> *"It may be that the Big Bang was the beginning. As Roger [Penrose] has emphasized for many years now, the observed fact is that our early universe is very, very special. If you took all the different ways you could*

[15] Impey, Chris. *The Living Cosmos*. Page 250. New York. Oxford University Press. 2001. Print.

> *arrange all the photons and particles of matter in the universe, the one that we actually have for our observed Big Bang is extraordinarily unlikely...that's these ten to the ten to the one-twenty-something unlikely and so either that was a tremendous accident and we got lucky but nobody believes that, or there's a reason why."*[16]

However, that is precisely the choice we have. Either the universe was created by accident, or it was created on purpose. A "planned accident" is merely a plan concocted to look like an accident. Carroll is an atheist, so he can't really believe there's a reason why the universe exists and remain logically consistent. He's saying he doesn't believe in the kind of luck necessary for an unplanned universe while steadfastly refusing to believe in the alternative, a planned and deliberate universe. To be perfectly honest, I'm not exactly sure what he's trying to say. The unplanned universe is atheistic by definition and produced by a remarkable coincidence of fortuitous events that must have succeeded, otherwise we would not be here. Conversely, a planned universe is not an accident. It is a deliberate construction.

From Frank Turek's website *CrossExamined* comes this analysis of the fine-tuning argument:

> *"The ratio of electrons to protons must be finely balanced to a degree of one part in 10^37. If this fundamental constant were to be any larger or smaller than this, the electromagnetism would dominate gravity, preventing the formation of galaxies, stars, and planets. Again, life would not be possible. The ratio of the electromagnetic force to gravity must be finely balanced to a degree of one part in 10^40. If this value were to be increased slightly, all stars would be*

[16] Carroll, Sean. "Big Bang Creation Myths", The Institute of Art and Ideas, January 31, 2019. https://www.youtube.com/watch?v=7HES3bPNAsA.

> *at least 40% more massive than our Sun. This would mean that stellar burning would be too brief and too uneven to support complex life. If this value were to be decreased slightly, all stars would be at least 20% less massive than the sun. This would render them incapable of producing heavy elements. The rate at which the universe expands must be finely tuned to one part in 10^55. If the universe expanded too fast, the matter would expand too quickly for the formation of stars, planets, and galaxies. If the universe expanded too slowly, the universe would quickly collapse before the formation of stars. The mass density of the universe is finely balanced to permit life to a degree of one part in 10^59. If the universe were slightly more massive, an overabundance of deuterium from the Big Bang would cause stars to burn too rapidly for the formation of complex life. If the universe were slightly less massive, an insufficiency of helium would result in a shortage of the heavy elements — again, resulting in no life."*[17]

Everyone is saying the same thing, more or less. The universe appears to be fine-tuned with incredible precision, and the evidence for fine-tuning is ubiquitous. The universe is fine-tuned. Our solar system is fine-tuned. A single-celled organism is fine-tuned. Human beings are fine-tuned.

Atheists typically look at the evidence of perceived flaws in living organisms and declare that this somehow counts as evidence against fine-tuning. The vas deferens appears to take a detour, so God is assumed to be a poor designer. The eye isn't as well made as a human would make it, because each eye has a blind spot. My response to those complaints is to ask, but where

[17] Unknown. "The Argument from Cosmic Fine-Tuning", CrossExamined, March 21, 2012, https://crossexamined.org/the-argument-from-cosmic-fine-tuning/

is the superior artificial eye? What if there might be a reason for the vas deferens to take a short detour? Things may be the way they are because a confluence of serendipitous coincidences culminated in an organism beginning to exist, or organisms may have been created by God. Why can't things be the way they are because of some need we humans don't comprehend? For example, whales have been claimed to have vestigial pelvic bones as remnants of their ancestors' terrestrial lives, but researchers at Harvard now say the pelvic bones are used when whales produce offspring.[18] Because we are human, we tend to make assumptions, and quite often, the foundation for our assumptions is wrong. Scientists originally thought that pelvic bones in whales served no purpose after they imagined that whales were once terrestrial animals and had legs.

However, the number of physiological changes that a whale would need to undergo to transform from a land-based animal to an aquatic creature have been estimated to exceed more than two hundred thousand successful mutations. It has been speculated that three unsuccessful mutations should be expected to result in the death of the organism in question. There is no reason to assume whales could have evolved from terrestrial animals, but the belief remains popular among evolutionary biologists.

A website calling itself Planet Curious has an interesting video about the problem of cosmological fine-tuning:

> *"The fine-tuning problem is a concept that arises in the context of the universe's physical constants and conditions. It refers to the observation that the fundamental parameters of the universe, such as the strength of gravity, the mass of elementary particles,*

---

[18] Ruell, Peter. "Status shift for whale pelvic bones", The Harvard Gazette, October 28, 2014.
https://news.harvard.edu/gazette/story/2014/10/status-shift-for-whale-pelvic-bones/

and the electromagnetic force appears to be precisely set in a way that allows planets such as ours to form and life like ours to exist because if these values were even slightly different, the universe would be a very different place. Changing the physical constants would play a crucial role in determining the behavior of atoms and molecules. It would therefore also affect chemistry. A slight change in some of the physical constants could lead to significant differences in chemical properties, potentially rendering the formation of complex molecules impossible. It would also alter nuclear reactions that power stars, which would lead to stars with different lifetimes, disturb the nuclear synthesis, and the process would therefore contain different abundances of elements. Any changes would lead to a completely unstable universe with vastly different cosmological scenarios, including a universe that collapses or expands too quickly. Proponents of the fine-tuning argument assert that the remarkable precision and delicate balance of the physical constants that govern our universe which makes the conditions necessary for life highly unlikely to have occurred randomly. They argue that this points toward the existence of a purposeful and intelligent designer who orchestrated the conditions for our universe and therefore also life. It suggests that the fine-tuning of the universe is evidence of an intelligent designer or creator commonly referred to as God." [19]

Either the evidence supports a belief in fine-tuning, or it doesn't.

---

[19] Unknown. "The Fine-Tuning Argument Debunked in 12 Minutes", Planet Curious, January 24, 2025. https://www.youtube.com/watch?v=KjRkOzKyoIw

Arguments Against Fine-Tuning

The evidence and arguments against fine-tuning aren't particularly good, quite frankly. They rely on absurd speculation and useless hypotheticals. No credible expert disputes that even the slightest change to any of the identified cosmological values would cause life and probably the universe to fail. The very best argument against fine-tuning is that IF all the cosmological variables were allowed to change dramatically, we can't be sure what the result might be—hardly an effective claim that our universe is not fine-tuned. A weak hypothetical is not a valid counterargument. It's merely a diversion. The fact remains that the universe capable of supporting life is fine-tuned, and a slight change to any of the identified cosmological values would cause life and our universe to fail. Alternate potential universes created by dramatically changing all the cosmological factors at the same time remain an untestable hypothetical solution.

In a podcast interview, physicist Brian Greene said:

> *"What we try to do is write down equations that will give us some insight into how it could be that there's a process that starts in the Big Bang and results in the things we are now experiencing. Now, are there certain things that might be pure coincidence? Like, could it be pure coincidence that the gravitational constant and the electromagnetic force have just the right values to yield the qualities that allow us to exist? It could be coincidental. It could be really causative. It could have been that if the values had been different, they would have caused something else, and that something else, if it had consciousness, might be in the same quandary, saying 'why were the constants just the right values to allow me to exist?' And yet they'd be different than the constants that we have. And so, you have to be careful about seeing things and being*

> *impressed at their uniqueness when they might not be unique, and impressed at how they are designed when they might not be designed. They simply may be what they are, and you are the result of those features. They weren't set in order to get you to exist. They were set, maybe by some random process, and you are the output of that random process. What we do as conscious beings, we tend to look at our environment and try to explain it in terms that somehow give us a narrative, a story, and we love stories that somehow have a purpose. But these stories may not have a purpose. It may be just random events with random numbers that yield based on their structure certain outputs, and we may be the output of those random qualities."*[20]

Have you ever tried to read a story without a point or a purpose? It's bad enough to read a story that never seems to have a point or go anywhere—how could anyone read a "story" that wasn't told with coherent sentences written in a native language? What about "stories" composed in complete gibberish? How would you interpret this next sentence?

> *SDegem xwhhhaztx cIm ddgiddd $h7efrbez?*

Utter gibberish, correct? Next, copy every odd-numbered letter in the sentence onto a piece of paper and read the result.

Is the sentence still gibberish? No, it is not. Assuming neither of us made a mistake in transcription, the sentence should read: See what I did here? The sentence above represents raw data. What you wrote down represents information. You represent a computer. You processed raw data and turned it into information. Order doesn't come from chaos without the

[20] "Does Fine-Tuning Point to God? — Brian Greene", Alex O'Connor, April 1, 2025, https://www.youtube.com/watch?v=2pkWs4QtP2Y

direction of an intelligent mind, much like information doesn't come from raw data without an intelligent mind directing how the data should be processed. Physicist Sean Carroll has a strong bias toward atheism, so he argues the fine-tuning argument is a terrible argument, even while conceding that even the slightest changes in the cosmological factors that determine the composition of the universe would cause life as we know it to not exist. Carroll's argument is that we don't know what a hypothetical universe might look like if the cosmological factors were to change significantly, but that is irrelevant to the existence of this universe. Carroll says that life as we know it would not exist, but that doesn't mean that no life would exist, and he claims that God wouldn't need to fine-tune the universe to create life.

When I read a well-known public figure like Dr. Carroll make such provocative claims, I am naturally tempted to respond. However, I've tried that once, with a less than satisfactory result. Dr. Carroll had appeared with a panel advocating for atheism, and he claimed to know what happens when we die (which is nothing). Because Dr. Carroll is an atheist, I have assumed that his answer would be nothing. So, I wrote Dr. Carroll to ask why he was so confident in claiming knowledge about what he thought he knew and asked if he was familiar with certain evidence that flatly contradicts that typical atheist worldview. Dr. Carroll replied that he would not examine any evidence he believed would contradict the laws of physics. How disappointing that is, to discover that a famous scientist you admired turns out to have a closed mind, screening evidence according to his confirmation bias.

First, we can ignore the hypothetical universe that we do not have and focus our attention on the actual universe that we do have. Is this universe fine-tuned? Most experts, including Dr. Carroll, appear to agree our universe appears to be fine-tuned in the sense that even the slightest change to any of the cosmological factors would have a disastrous effect on the universe, causing life and probably the universe itself to not exist.

Sean Carroll's argument against fine-tuning is specious. He asserts the following:

1. Carroll concedes that if you changed the parameters of nature by the slightest bit, the local conditions would change by a lot, and the universe would fail. He rejects the argument that changing all the parameters a lot should also be expected to fail.

2. Dr. Carroll argues that God doesn't need to "fine-tune" anything. He could have created a universe without rules.

3. He claims the fine-tuning parameters could disappear as our understanding of the universe increases. Carroll claims the expansion rate of the early universe was fine-tuned to within one part in 10 to the 60th power, but suggests that if you work through the equations of general relativity with the right derivations, the probability becomes 1.

4. The multiverse eliminates concerns about fine-tuning in this universe.

5. Even if fine-tuning is true, theism fails to explain it.

A rebuttal to Dr. Carroll's argument:

1. Useless hypothetical. We know our universe is special because scientists have identified a set of fine-tuned values. Changing those values even slightly would cause the universe to fail, and humans (observers) would not exist. Changing each of those cosmological values by a lot would almost certainly mean that humans did not exist.

We'll never know because we already have our fine-tuned universe.

2. Another incoherent argument and useless hypothetical. The word 'need' is very strange to use in this context. Should we expect the universe to have rules but expect God to demonstrate His power and might by violating His own rules? Because we know very little about the nature of God, it is patently foolish for humans to place limitations on His power. Carroll seems to be implying that God could have created a universe without natural laws. Why assume God would want to create a universe with no rules? In a universe with no rules, what would a miracle be?

3. I'm not going to argue mathematics with Dr. Carroll. A man must know his own limitations. I would simply point out that a probability of 100 percent represents absolute certainty, which only a foolish person would claim. I'm confused—is Dr. Carroll suggesting that our universe had no choice but to exist? A probability of 1 suggests absolute certainty.

4. Theism gives us God as the First Cause. The multiverse "solves" the problem of fine-tuning with a claim allowing for the creation and destruction of an infinite number of failed universes, which had a unique set of "incorrect" fine-tuned parameters that caused that universe to fail. However, the multiverse theory fails to speculate as to the actual cause of the Big Bang and fails to speculate about a reason why the Big Bang happened.

5. If the universe is fine-tuned (as it clearly appears to be), then God must be considered for the role of the Cosmic Architect, the One who adjusted all the knobs and dials perfectly to fine-tune this universe to be just right for human life. The alternative explanation is luck, or

random chance, and sufficient time to make such luck a bit more palatable.

During a debate against William Lane Craig, philosophy professor Peter Millican claimed,

> *"So, the fine-tuning argument is based on some apparent coincidences in physics that are genuinely intriguing, and I would concede this much to it. If there is a physically inexplicable coincidence of the fundamental constants of nature whose values have to be precisely tuned within a wide range of otherwise available possibilities to make a complex universe possible, then this constitutes a phenomenon which very naturally invites explanation in terms of a cosmic scale designer. But I also see good reason for caution and modesty here. First, we must beware our natural tendency to see purpose in things too readily. This can make us bad judges of what is an impressive coincidence and what is not. Secondly, we only have experience of this universe, so even making sense of cosmic probabilistic claims is problematic. Our experiments and observations have a sample size of one and we really have no idea of the range of possible alternative scenarios. Thirdly, we can't rule out that some deeper explanation will be found for any apparent coincidences. For example, Guth's theory of cosmic inflation can explain something in some of them. There might be multiple universes such as a world ensemble, an evolving sequence or a complex of bubble universes each with its own Big Bang and different conditions. Then it would be no surprise at all that observers have evolved only in those universes that are conducive to life. Fourthly, the physics on which this entire argument is based is young, incomplete, and radically imperfect. It's less than 100 years since general relativity and quantum mechanics, our best theories of the large and the small were*

developed and we still have no way of reconciling these with each other. String theory attempts this and speculates that the universe has eleven dimensions in order to facilitate it, but that still has a very long way to go and is widely regarded with amused skepticism by experimental physicists. "[21]

We haven't yet had the opportunity to discuss Guth's theory of cosmic inflation. That comes next, as the second of our seven miracles. Millican seems to be saying that it's not only impossible to know whether probabilistic claims are valid due to the sample size, while conceding the fine-tuning argument invites the introduction of a "cosmic scale" intelligent designer into consideration. It's an honest admission.

The Multiverse Hypothesis

Our created universe shouldn't be here, but it is. The universe shouldn't be capable of supporting life, but it does. The life the universe supports shouldn't be intelligent if it were created by unintelligent processes, yet intelligence exists. A lot of things are that ought not to be. Remember, multiverse hypotheses exist solely to mitigate the probability problem posed by our extraordinarily improbable universe. The multiverse doesn't solve the cause problem or the problems of intelligence and consciousness. Physicist Brian Greene explains why multiverse hypotheses exist:

"One explanation simply would be there are many many many many universes in which the parameters of all different values across those many universes and in most of those universes the conditions are not amenable to living systems like human beings, but in

---

[21] Unknown. "The Fine-Tuning Argument Debunked in 12 Minutes", Planet Curious, January 24, 2025. https://www.youtube.com/watch?v=KjRkOzKyoIw

> *one of those or a small number of those universes the conditions are amenable and of course we're in one of those universes because we could not exist in any of the others because the conditions are not correct.*[22]

No gambler in his right mind would bet on the success of an unplanned, fine-tuned, perfectly inflated anthropic universe populated with complex living organisms capable of contemplating their own existence, and yet somehow here we are. Proposing alternate explanations for failure doesn't really explain success. Okay, so how did we get here? Why do we exist? The universe is fine-tuned to exist in its current form and fine-tuned for complex life on this planet. Greene continues,

> *"It's [fine-tuning evidence] just a fairly complicated problem that you don't want to dismiss too quickly by saying it all goes away unless it's precisely as we've seen, but it [the probability problem] is the motivation for the multiverse." In the same interview, Greene later explains, "It is a natural way of avoiding the problem that you're making reference to—the specialness of this universe goes away if it's one of a grand collection of possibilities."*[23]

Why should we rely on a hypothesis that only exists to solve a probability problem and has no evidence to support it? Especially since the multiverse can only reduce the improbability of this universe by making it one of many "universes" (a contradictory term to be sure, given that the prefix "uni" means one, as in the only one universe that we

---

[22] Greene, Brian. "Does Fine-Tuning Point to God? — Brian Greene", Alex O'Connor, April 1, 2025, https://www.youtube.com/watch?v=2pkWs4QtP2Y

[23] Greene, Brian. "Does Fine-Tuning Point to God? — Brian Greene", Alex O'Connor, April 1, 2025, https://www.youtube.com/watch?v=2pkWs4QtP2Y

know exists. Greene acknowledges the real reason that multiverse hypotheses have been conceived—to make the idea of an unplanned universe slightly more plausible rather than completely unbelievable. But even if the multiverse did exist, it would provide no explanation for the catalyst that caused the Big Bang. It merely addresses the improbability of the unplanned universe. Greene says,

> *"What fires me up is the fact that a collection of particles called a human brain which is all that we are, collections of particles, can coalesce through an evolutionary dynamics to yield a structure that can think and feel and love and emote and create and illuminate and figure out quantum mechanics and figure out general relativity and come up with the idea of a multiverse and describe black holes and predict the magnetic moment of the electron—that to me is the amazing thing—that matter not infused with any divine force, not in my view structured by some divine plan, can through the bare laws of physics come together and do what it does."*[24]

I didn't intend to pick on Greene, but he did slip up and admit the real reason the multiverse exists—to counter the probability problem that a planned universe doesn't have.

Cosmic inflation happened either during or immediately following the Big Bang, but it has its own separate set of calculated probabilities independent of the Big Bang. Cosmic inflation is considered a separate and unique miracle worthy of its own section of the book. The multiverse might give us a potential explanation for the origin of the universe, but it can give us nothing in the way of an explanation for cosmic

---

[24] Greene, Brian. "Does Fine-Tuning Point to God? — Brian Greene", Alex O'Connor, April 1, 2025, https://www.youtube.com/watch?v=2pkWs4QtP2Y

inflation. Nor does the multiverse offer us a candidate for First Cause.

ANALYSIS OF THE ORIGIN OF THE UNIVERSE

Planned Versus Unplanned

There is a number that represents the probability our universe was the unplanned effect of some unknown cause. It is not my number. It is a number produced by experts in their respective fields who performed difficult calculations to estimate the probability of the confluence of events that made the origin of our universe possible. The Nobel laureate Sir Roger Penrose, an atheist, performed many of these calculations and determined that our existing universe is enormously improbable. If the unplanned universe is extremely improbable, then simple logic dictates that a planned universe is conversely highly likely. The cosmological factors underlying fine-tuning would not be any different depending on whether the universe is planned or unplanned. The unplanned universe must resolve all the calculated improbabilities associated with fine-tuning. The planned universe scenario simply invokes an intelligent planner to increase the probability of success close to 100 percent.

If we simply allow for the possibility of a planner, the probability that our fine-tuned universe could pop into existence out of nothing rises to virtually 100 percent. Technically speaking, the probability can't literally be 100 percent because 100 percent represents absolute certainty on the probability scale, but the probability of a planned universe does rise to nearly 100 percent if we simply take the inverse probability of the unplanned universe. In other words, if the probability of an unplanned universe is .0000001 percent, the probability of a planned universe would be .9999999 percent. The odds against our fine-tuned universe popping into existence without any help are astronomically low...the actual value is not one in one hundred, one in a million, one in a billion, or even one in a trillion. The odds against success are what I call a silly number that involves exponents and lots of zeroes—quite a few more than the seven zeroes used in the example.

I remember reading that Roger Penrose had performed calculations based on the work of Martin Rees and declared the probability of our fine-tuned universe popping into existence due to random chance was something along the lines of 1 in 10^300. I've also seen an interview with Penrose where he seemed to be saying that same prediction of improbability was 1 in 10 in 10^124. Frankly, I'm not entirely sure which is the smaller number, although "1 in 10" seems to be at least an order of magnitude larger than "1 in 10 in 10." In my opinion, once you get past 1 in 100 and 1 in a million, you're talking about fairly unreasonable probabilities.

Prior to the Big Bang

The consensus opinion is that before the origin of this universe, *something* must have existed for the simple reason it's always easier to believe something (the universe) came from something. According to physicist Lawrence Krauss, it was something called quantum foam. Other physicists, including Sean Carroll, have proposed the existence of a static, eternal parent universe and that our specific universe is a child universe that succeeded where an unknown number of child universes failed, a multiverse hypothesis. Roger Penrose thinks that our universe might expand and contract in cycles. Nobody knows what might have existed prior to the Big Bang. It will never be possible to know. All that we know with confidence is the fact that our universe has not always existed in its current form. It has been said that if we could reverse course throughout history and head backward toward the Big Bang, our universe would gradually condense to the point where the entire universe was smaller than the tip of a pin.

Unfortunately, we can only go back so far in time before we reach an impassable boundary at the moment where space and time began to exist. Science podcaster Arvin Ash explained,

> *"You might think, wait a minute! We have all these models that describe what happened fractions of a second after the Big Bang so well. Why can't we just turn the clock back a little further back and figure out what happened just a tiny fraction of a second before that? What is the problem? The problem is that our current understanding of physics breaks down at the singularity, the moment that predates the Big Bang. Beyond that singularity, all our theories tell us nothing. The laws of physics as we know them cease to apply. It's an informational boundary."*[25]

We can speculate and make any sort of guess about what might have existed prior to our universe, no matter how crazy it sounds, simply because no one will ever be able to prove the claim wrong. What is quantum foam? Who knows? Has anyone ever seen quantum foam? We might be able to mathematically prove the *possibility* of quantum foam, but as the old expression goes, seeing is believing. I'll believe quantum foam exists when I can see it or at least feel it. It's difficult to believe that something abstract and unintelligent could create this universe *ex nihilo* with the degree of precision required.

There are three possibilities regarding what might have existed prior to the Big Bang. These three options are:

1. **Literally Nothing**. The universe was created *ex nihilo,* from nothing.

---

[25] Ash, Arvin. "What was there before the Big Bang? Three Good Hypotheses!", Arvin Ash, April 13, 2024, https://www.youtube.com/watch?v=QAuf97BhgRY

2. **Cyclic Universe**. The universe has always existed, expanding and contracting, sort of like an accordion. Advantage: it avoids the need to explain the true origin of our universe, and Roger Penrose likes it. Disadvantage: It doesn't explain the cause of the Big Bang that started THIS universe. It merely offers a potential explanation for the material that became matter after the Big Bang. Also, observations of the universe indicate the universe continues to expand at an accelerated pace, which would appear to contradict the idea the universe will begin to slow back down and eventually contract at some point.

3. **Multiverse**. An unknown and perhaps even infinite number of universes all popped into existence simultaneously. Most of the other universes failed, but ours succeeded. Advantage: it would appear to solve the probability problem in our fine-tuned universe by introducing numerous other failed universes, thereby literally improving the odds of success. Disadvantage: Again, it doesn't explain the cause of the Big Bang that started this particular universe.

Even the evangelist for atheism himself, Richard Dawkins, has conceded the evidence for fine-tuning exists:

> *"It does seem to be the case that physicists have measured half a dozen or so fundamental constants—things like the gravitational constant which are very critical in the sense that if they were ever so slightly different, the universe as we know it wouldn't come into being. There wouldn't be galaxies. There wouldn't be stars. There wouldn't be chemistry. There wouldn't be planets. There wouldn't be us. There wouldn't be evolution. And so, it can be argued that these constants are very finely tuned, and if they weren't finely tuned, we wouldn't be here. Well, we are here, so evidently, they were finely tuned, and there's the temptation to*

> *think they were tuned by some sort of intelligent being, but that is hopeless because that doesn't explain anything. It just pushes the problem back to how the intelligence itself got tuned."*[26]

Dawkins likes circular arguments, but he is asking the wrong question. The right question is, how did intelligence itself come to exist? If our intelligence comes from an unintelligent source, can we trust it? A planned universe is an intelligent universe. An unplanned universe needs to coherently explain the origin of intelligence. Without a successful Big Bang and successful cosmic inflation, life could not exist. And if life does not exist, it cannot evolve. Dawkins fails to acknowledge the logical alternative to his bottom-up explanation for life, the universe, and everything—a supernatural form of intelligence that exists outside of space and time. Some people might like to call this form of intelligence "God." No matter what we might call the Creator, an entity that created our space and time must by definition exist outside of our space and time and would be considered eternal by definition.

St. Thomas Aquinas eloquently argued that God created both matter and time when the universe was created. Indeed, the moment of the Big Bang is sometimes referred to as ***t=0***, where t represents time itself. In other words, ***t=0*** represents the moment time began. Time prior to ***t=0*** is undefined, for good reason—prior to the Big Bang, there would have been no way to measure time.

---

[26] Dawkins, Richard. "God Doesn't Solve the Fine-Tuning Problem", The Poetry of Reality with Richard Dawkins, September 6, 2023. https://www.youtube.com/shorts/40oTwPiGg3I

The Goldilocks Enigma

Physicist Brandon Carter asked a significant question: "Why are there laws of physics?" Why indeed? For that matter, why does logic exist? Why does anything exist?

Cosmologist Paul Davies wrote a book titled *The Goldilocks Enigma: Why is the universe just right for life?* In the book, Davies writes,

> *"Many scientists who are struggling to construct a fully comprehensive theory of the physical universe openly admit that part of the motivation is to finally get rid of God, whom they view as a dangerous and infantile delusion."*[27]

Later in the book, Davies also writes,

> *"I described the singularity in the movie-in-reverse account as 'the vanishing point' of the universe. But why did it have to vanish? Could the singularity not have just sat there? In forward-time description, there would be a singularity—think of a point of infinite density if you like, a structureless, size-less cosmic egg—existing for all eternity, when suddenly it went 'bang.' In that case, what came before the Big Bang would no longer be nothing, it would be "a singularity." Some popular accounts of the origin of the universe promulgate that dubious notion. However, it won't do. The theory of relativity links space and time together to form a unified spacetime. You can't have time without space, or space without time, so if space cannot go back through the Big Bang singularity, then neither can time. This conclusion*

[27] Davies, Paul. *The Goldilocks Enigma: Why is the Universe Just Right for Life?* Page 15. Boston. Mariner Books. 2008. Print.

> *carries a momentous implication. If the universe was bounded by a past singularity, then the Big Bang was not just the origin of space, but the origin of time, too. To repeat: time itself began with the Big Bang. This neatly disposes of the awkward question of what happened before the Big Bang."*[28]

Paul Davies is a highly respected subject matter expert. If Davies is correct and the Big Bang was the origin of time, space, and matter (as Kent Hovind also suggested), then nothing existed or happened in this universe prior to the Big Bang. Time, space, and matter all began with the Big Bang. When asked if the Big Bang theory could be mistaken, the well-regarded physicist Sean Carroll replied,

> *"We first have to say what we mean by the Big Bang theory because this phrase is meant in two very different contexts, right? We all know the universe is expanding, so if you run the clock backwards, run the film through the past, 14 billion years ago, it was in a hot, dense state. And we have something called the Big Bang model of cosmology, which is simply the statement that 14 billion years ago, the universe was in a hot, dense state. It expanded and cooled, and went from being very smooth to relatively lumpy, with all these stars and galaxies and so forth. That's the Big Bang model; it is true. There is no point in doubting the Big Bang model."*[29]

It isn't so much doubting the Big Bang model as it is quibbling about the interpretation of events that we can neither explain nor understand, and mostly what might have preceded the hot,

---

[28] Ibid. Page 222.

[29] Carroll, Sean. "Big Bang Creation Myths", The Institute of Art and Ideas, January 31, 2019. https://www.youtube.com/watch?v=7HES3bPNAsA.

dense state of the very early universe. Our existing (or the current) universe is awfully large. However, from my own perspective, the Earth is awfully large, and I am very small by comparison. But compared to an ant, I am huge. As comedian Steven Wright put it, "It's a small world...but I wouldn't want to paint it."

It is also extraordinarily difficult to imagine the entire cosmos could have once fit upon the head of a pin, yet science tells us that was the case. Roger Penrose believes the cyclic universe is possible, and Sean Carroll disputes fine-tuning, but they both still agree the Big Bang event occurred. Carroll added,

> *"We know exactly what the universe was doing one minute after the very beginning of the Big Bang. From one minute after to 14 billion years after, we understand. That first minute is a little bit up for grabs. So, classical general relativity, the theory that Einstein gave us for space and time, would say, according to Roger (Penrose) and Stephen Hawking, that at that moment where T equals zero at the very beginning, there was a singularity, but there's also this thing called quantum mechanics which gets in the way which is not part of general relativity. So, if you want to say the Big Bang event, the Big Bang moment, the beginning of everything, we don't know whether that is right or not...it's a moment in time, the Big Bang, not a place in space. It's not an explosion into preexisting space. It's the beginning of everything. It's the moment before which there were no other moments. That's the model, and the question is, is that model right?"*[30]

Has the argument about the origin of the universe truly been reduced to the very first minute? Is that all that's left to debate?

---

[30] Carroll, Sean. "Big Bang Creation Myths", The Institute of Art and Ideas, January 31, 2019. https://www.youtube.com/watch?v=7HES3bPNAsA.

I don't think so. The multiverse doesn't explain cosmic inflation. It doesn't explain abiogenesis. It doesn't explain consciousness. It doesn't do anything but offer scientism an excuse to consider alternatives to our creator God. In an interview, Stephen Meyer said,

> *"What the James Webb telescope is able to do, in fact, what it was constructed to do, was to detect extremely long wavelength radiation, stuff that's outside the visible range. I call it the uber redshifted—it's actually in the infrared range, is the more accurate physics term. So, it's looking for very long wavelength radiation coming from galaxies that are very, very far out there. Now, why would it be looking for that? Well, because if the universe is expanding as we would expect based on the Big Bang theory, then the radiation coming from things very, very far out in space and therefore very far back in time should be very stretched out, more stretched out than stuff closer at hand. So, the James Webb was constructed in hopes of detecting that type of radiation if it existed. It's not assuming that it necessarily would, but it would be a way of confirming the expansion of the universe has been going on for a very long time...What they were in fact able to detect from these very ancient, very distant galaxies was super redshifted radiation. Uber redshifted stuff out in the infrared. On the basis of that, they were able to synthesize images of these very, very distant, remote galaxies. Now, the very fact they were able to do that confirms that you have what you'd expect on the basis of the Big Bang theory, that the amount of redshift that you would expect to be present if in fact the galaxies had been expanding throughout that vast stretch of time was in fact present and was detected. Now that didn't get reported. The whole focus was on the fact there were galaxies that were more mature and more of them early on than we expected based on our theories of galaxy formation.*

> *And so, those are anomalies that need to be addressed and have not been explained, as I understand it. Maybe the astrophysicists have made more progress on that in even recent days. But the basic picture of an expanding universe outward from a beginning has not been undermined but rather confirmed in a very dramatic way at very great distance at a very far look-back time, way, way back in time. So, I think it's a rather dramatic confirmation, but there have been many others—the cosmic background radiation that was discovered in 1965, the COBE radiation that George Smoot discovered in the Nineties. So, there's this pattern of confirming evidence of this basic picture of an expanding universe outward from a beginning in observational astronomy from the Twenties right up till now. And so that gives us a good reason to think, as best we can tell, the universe had a beginning."*[31]

That should effectively end the debate as to whether our universe had a true origin, but of course it won't because the debate has such theistic implications and there will always be those who resist the concept of God with both tooth and nail. The cyclic universe and multiverse try to avoid the problem of the universe having an origin by separating the origin of the universe from the Big Bang, but neither resolves the problems created by moving forward from the Big Bang—problems including such as why was the universe so exquisitely fine-tuned? Why does the universe contain the perfect composition of materials to form life? Why does life exist? How important is water to the existence of life? These existential questions are important and interesting. They can be asked, but they will never be answered, unless one day we have the opportunity to

---

[31] Meyer, Stephen. "Intelligent Design Expert on the Big Bang and the James Webb Telescope", PowerfulJRE, July 13, 2023. https://www.youtube.com/watch?v=tb1Ubw1Iu5w

ask God directly. By then, our physical bodies will have failed, and they will become food for worms.

Potential First Causes

In a planned universe, the uncaused First Cause is a creator God. In an unplanned universe, the candidates for First Cause appear to be unknown. What might be the candidates for First Cause? Time? Luck? The laws of physics? Repulsive gravity? The Big Bang is simply an effect. We need to consider the cause. Physicist Brian Greene was asked what caused the Big Bang and he said, "Why would something smaller than the head of a pin become everything that we see in the cosmos?"

An excellent question. Greene replied,

> *"So, there are ideas for the answer to that question. Look, all of this is tentative because it's very hard to do measurements that go all the way back to the beginning. We have astronomical observations that we need to be sure are compatible with the predictions of our theories and so forth, so we, as good scientists, do what needs to be done to try to test these ideas. But the idea that I think most physicists or cosmologists buy into at the moment is that gravity can have two manifestations. The usual form of gravity that you and I know about is the attractive version. You drop something toward the Earth, and it moves downward because the Earth and the object pull on each other. That's the ordinary gravity we experience every day of our lives. But Einstein's equations actually allow gravity to also be repulsive. It can push outward as opposed to just pulling inward. And this is something that we have never experienced because the gravity created by a rocky object like the Earth is always the attractive variety. The gravity created by the sun—again, a compact object is always the attractive variety. But Einstein's math shows that if you don't*

have a rocky object that's isolated in space, but rather energy that is uniformly spread through a region of space—that that kind of entity yields repulsive gravity. Why is that important to your question? If the very early universe—that little, tiny head of a pin you're talking about—if it was filled with a uniform bath of this energy, we call it the inflaton field. The name doesn't matter, but if it were filled with that energy, it would have been subject to repulsive gravity. What does repulsive gravity do? Pushes everything apart. Causes everything to rush outward. So, the bang of the Big Bang might have been a spark of repulsive gravity operating within a tiny region of space that pushed everything apart."[32]

Let's see if I understood that correctly...gravity as we know, experience, and observe it is an attractive force that pulls objects toward each other, but during the brief fraction of a second while cosmic inflation occurred either during or immediately after the Big Bang, it acted as a repulsive force repelling objects (particles) from each other that briefly continued until inflation ended and gravity reversed. Is that correct? From where did this energy come? Alan Guth, the man who conceived of cosmic inflation, coined the name for the inflaton field to describe this hypothetical scalar field that existed immediately following the Big Bang. There were particles needed to form atoms and molecules.

Perhaps all the building blocks were there, but science doesn't seem to offer explanations for their existence.

---

[32] Greene, Brian. "What caused the Big Bang? Brian Greene explains", RoganOnRepeat, March 3, 2025. https://www.youtube.com/shorts/-oZW_68U_fA

Alternatives to the Big Bang

Acceptance of the Big Bang theory is widespread, but it is not universal. The sticking point is whether the Big Bang itself was also the true origin of the universe.

There are competing theories, such as Cyclic Conforming Cosmology and Big Bounce theory, both of which speculate that the universe could be eternal and had been contracting prior to this most recent state of expansion. It is unclear whether the strategy of avoiding the problems caused by the singularity creating this universe from nothing by making the universe eternal, can also resolve the problems created by fine-tuning and precision of the continuously expanding universe. It doesn't explain why this universe is compatible with complex life and the life that exists on this planet. It's as if everyone recognizes that *something* involved in the explanation for the universe itself must be eternal. Non-religious people appear to have assumed that the eternal entity must be the universe itself, while advocates for a planned universe will probably wish to credit the eternal God. It is my understanding that the argument isn't about whether redshift and the CMB are legitimate evidence for an expanding universe. All of the disagreement and focus is on what could have existed prior to the beginning of the current expansion. Even a previous universe offers no explanation for why *this* universe exists. A truly eternal universe would be the steady-state or unchanging universe.

Our universe is probably quite old, but it isn't eternal. Even if we assumed the universe is eternal, the moment the previous universe ceased contracting and the current universe began to expand is still the Big Bang. It is the exact moment our current universe began to expand—the only difference is that the expansion began as a little ball of material that wasn't matter that originally came from some unknown source, versus expansion that began as a little ball of material that was formed from a previous universe. The cyclic and bouncing universes try to get around the origin of matter problem by suggesting matter itself is eternal, but none of that explains why this iteration of expansion created the universe we currently inhabit.

Some very smart people advocate for the cyclic universe and bouncing universe because it does get rid of the singularity, but it doesn't explain why the universe would need to expand and contract or how the previous universe could collapse to a vanishing point and then virtually explode into becoming this universe we occupy at an accelerated but precise rate of expansion. Obviously, either God or the universe must be eternal. The problem with assuming the universe is eternal is that an eternal universe is not required to be intelligent, but there should be an intelligent explanation for intelligence to exist. Otherwise, we need to explain how intelligence came from a non-intelligent origin. If intelligence and our thoughts are merely the result of chemical reactions in a physical brain that is merely coding for protein, why should we bother to trust these thoughts?

Most of the debate is not about whether the Big Bang occurred, but about what might have existed before that moment when ***t=0***. However, acceptance of the Big Bang theory is still not quite universal. Science writer Eric Lerner has essentially launched a one-man campaign to advocate for the discredited steady state theory. Lerner's work has been cited in opinion pieces such as "Edwin Hubble's FALSE Law," written by David Rowland. The paper by Rowland is formatted to resemble a scientific paper and makes a number of sensational claims, such as "The Universe is not Expanding," allegedly based on measurements of the surface brightness of galaxies. In his conclusion, Rowland even accused Edwin Hubble of falsifying evidence to support Hubble's Law, saying,

> *"Hubble's law, the alleged definitive evidence supporting expansion theory, is fatally flawed. Edwin Hubble made the a priori assumption that galaxies are*

> *moving away from each other, then falsified evidence to support his foregone conclusion."* [33]

Wow! I suppose it is possible that Eric Lerner is this generation's Galileo and has the correct answer about the universe, even though virtually every other scientist who studies the origin of the universe disagrees with his conclusions. But it is also quite possible that Lerner isn't Galileo. He's just another science writer motivated to make a name for himself. Everyone disagrees with him because he was egregiously wrong to accuse Edwin Hubble of falsifying evidence and wrong to insist the steady-state universe is correct when the Big Bang and cosmic inflation appear to give us the best explanation for the universe we have today.

The bouncing or cyclic universes only provide an alternative explanation for the origin of matter by eliminating the singularity. They resolve the problem of nothing simply by insisting *something* has always existed, and that something could conceivably be a previous universe. However, that proposed solution doesn't explain why this current universe was so precise in its formation or why the universe contains intelligent life.

Kalam Cosmological Argument

Christian philosopher William Lane Craig has re-popularized the Kalam Cosmological Argument as part of his argument for the existence of God. The argument is straightforward:

1. Whatever begins to exist has a cause.

2. The universe began to exist.

---

33 Rowland, D. "Edwin Hubble's FALSE Law", Open Scientific Publishers, 2022, https://www.ospublishers.com/Edwin-Hubble%E2%80%99s-FALSE-Law.html

3. Therefore, the universe has a cause.

Critics of the argument like to say that the true origin of the universe was not the exact moment of the Big Bang, but that counterargument is irrelevant. The argument is that the universe began to exist, period, not that the precise moment must be called the Big Bang. If cosmic inflation actually preceded the Big Bang, then we should expect that the true origin of the universe was at least slightly before the moment of the Big Bang itself. We know this universe has not always existed because when the Big Bang did occur, there were no existing stars or planets, no galaxies or living organisms. The argument we don't know the precise moment when the universe began to exist is primarily a semantic one.

The Big Bang occurred, and the universe had a true origin.

Problems Created by the Origin of the Universe

There are at least four significant problems with the Big Bang theory describing the origin of our universe. These problems are:

1. The universe appears to have been created out of nothing.
2. The universe appears to be very precisely fine-tuned.
3. An unplanned Big Bang has no identifiable First Cause.
4. The universe must be stretched (cosmic inflation, a second miraculous event) for the Big Bang model to work.

As previously noted, prior to the Big Bang, no stars, planets, galaxies, or living organisms existed. No plants, no animals, no matter, no chemicals. Not even one single atom existed. Some physicists have since changed their tune and claim the universe existed in some capacity prior to the Big Bang, either as a previous universe, quantum foam, or "something" that doesn't exist in this universe. Scientists have studied the chemical composition of the universe and performed calculations to estimate the various forces and their effects on the universe as a whole. Sir Martin Rees wrote a book titled *Just Six Numbers* that identified six cosmological variables that even the slightest variations in their values would cause our universe to fail to exist. Although cosmic inflation is treated as a separate event in this book because it has a different probability estimate than the origin of the universe, inflation is very closely tied to the Big Bang, either beginning shortly before the Big Bang and ending with the Big Bang, or beginning and ending in a fraction of a second immediately after the Big Bang.

There is no identifiable First Cause for an unplanned Big Bang, meaning there is no explanation for why the Big Bang occurred. What event triggered the beginning of the universe?

Arvin Ash said,

> *"Today, the evidence for the Big Bang model is so strong that virtually no scientist seriously disputes that it's an accurate description of the history of our observable universe. But some observations leading up to the 1970s had found some mysteries that the model in its current form could not account for. In order to develop into the universe that we see today, the early universe had to have some very specific properties, some of which seemed implausible with the early Big Bang model. Specifically, the questions left unanswered by the Big Bang were: why was the early universe so uniform? Why is the universe so close to being geometrically flat? Why do we not see any*

> *magnetic monopoles, which theoretically should exist?"*[34]

While we're asking questions, what is a magnetic monopole, and why should we assume they should exist but don't?

Katie Mack, Hawking Chair of Cosmology and Science Communication for the Perimeter Institute, said:

> *"There's a popular story about the Big Bang that you've probably heard before. The universe began as a singularity, a point of infinite density and zero size that appeared from nothing. That singularity contained all of space and time, and then it expanded, and everything in the universe came from that single point."*[35]

The evidence seems rather conclusive that the Big Bang theory describes the origin of our universe, or the moment immediately after the origin of the universe, which also might have been when cosmic inflation occurred.

CONCLUSION

We can be reasonably sure that the universe has not always existed. We know (because scientists tell us) that there was once a point in time when no stars, planets, or life existed. Scientists call the moment that space, time, and matter began to exist the Big Bang. We may not be able to say for certain what existed before the Big Bang, but we do know that whatever it was, it was

---

34 Ash, Arvin. "Cosmic Inflation: The Solution to the Big Bang Theory and the Universe", Arvin Ash, February 26, 2022, https://www.youtube.com/watch?v=q5dy-NtVeF0&t=202s

35 Mack, Katie. "Cosmic Inflation Explained | Cosmology 101 Episode 6", Perimeter Institute for Theoretical Physics, August 9, 2024, https://www.youtube.com/watch?v=RHBlJHL7cBs

not this anthropic universe. It was probably nothing. Literally. Nothing except God in Heaven.

The Bible offers God as the First Cause of the Big Bang. The first chapter of Genesis, the very first verse, says, "In the beginning God created the heavens and the earth."

The Bible also tells us how God created the universe. By speaking. God spoke and said, "Let there be light." And there was light.

MIRACLE 2
THE EXPANSION OF THE EARLY UNIVERSE

INTRODUCTION

The universe is said to be approximately 90 billion lightyears across, but it is allegedly only 13.8 billion years old. How can this be possible? Why the vast discrepancy? The most logical answer seems to be that the early universe briefly expanded at a much faster rate than it is expanding today, a scientific theory known as cosmic inflation. Cosmic inflation describes the very brief period of time that occurred immediately before or immediately after the Big Bang, where the universe expanded at a pace considerably faster than the speed of light. When inflation ended, the universe continued to expand, but at a much slower rate that continues today. The combined expansion of cosmic inflation plus the ongoing expansion of space appears to adequately explain the total size of the universe. There are (allegedly) at least three significant problems associated with the Big Bang theory: first, the universe appears to be flat (flatness or oldness problem), and second, the universe appears to be uniform no matter which direction we look (horizon problem). The third and final problem is the absence of theoretical particles called magnetic monopoles.

On his podcast that provides scientific education, Arvin Ash describes cosmic inflation as follows:

> *"Note that not all physicists agree that the beginning, that is, the point when time first started, t equals zero, is the same as the beginning of the Big Bang. Some physicists believe that the Big Bang happened after inflation, around 10^-12 seconds. Also note that some descriptions of inflation say something like the universe started out smaller than an atom and then expanded to the size of a grapefruit. You should keep in mind that these analogies can be misleading because they imply that the universe has an edge. It doesn't. There is only the universe and nothing else."*[36]

Our universe is described as homogeneous, meaning it is uniform in structure, and isotropic, meaning it appears to be identical when looking in every direction. There appears to be a uniform temperature across the universe, with a variance of less than a fraction of a degree. No matter which direction we choose to look in the night sky, the composition and temperature of the universe appear to be roughly the same. It seems logically reasonable to assume that if the universe had a specific starting point, it would have a cartological orientation. For the sake of argument, let's say the universe originated at some point north of our solar system. If we look to the north in the night sky with a powerful enough telescope, we should be able to see the glow from the CMB. However, if we look in the other direction, due south, we should not see the CMB, but we do. Apparently, this is because of cosmic inflation. Much like the agreement about the Big Bang theory, there doesn't seem to be vehement disagreement among physicists and cosmologists that cosmic inflation, or the rapid expansion of the early universe, happened either immediately before or immediately after the Big Bang singularity, but neither is there a universal consensus that inflation theory is correct. Paul Steinhardt, Avi Loeb, and

---

36 Ash, Arvin. "Cosmic Inflation: The Solution to the Big Bang Theory and the Universe", Arvin Ash, February 26, 2022, https://www.youtube.com/watch?v=q5dy-NtVeF0&t=202s

Anna Ijjas have argued that alternative explanations to cosmic inflation warrant further exploration.

Some believe the Big Bang happened, and then cosmic inflation began and ended fractions of a second later. Others believe the expansion of the early universe ended with the Big Bang. It would seem to make the most sense for inflation to follow the Big Bang because the Big Bang is the moment t equals zero, or the moment that space, time, and matter all began to exist, at least according to the quantum mechanics model of the Big Bang theory. In multiverse hypotheses and cyclic universe theory, space might not have been created by the Big Bang, but time and matter were. Immediately following (or a fraction of a second before) the Big Bang, the universe expanded at an incredibly accelerated pace for an extremely precise period. According to the experts, if inflation, which only lasted for a fraction of a second, had been longer or shorter, the universe we currently inhabit would not exist. The force that allegedly caused cosmic inflation is a special form of gravity called repulsive gravity that could only exist in theory when the universe was in this special state. Arvin Ash adds,

> *"The simplest version of the theory of inflation says the universe expanded exponentially fast, faster than the speed of light, near its earliest history—from what is believed to be the beginning of the universe, from about $10^{\wedge -36}$ seconds after the big bang to $10^{\wedge -32}$ seconds. During this time, it expanded by a factor of at least $10^{\wedge 78}$, going from being very small to being exponentially large in comparison."*[37]

---

37 Ash, Arvin. "Cosmic Inflation: The Solution to the Big Bang Theory and the Universe", Arvin Ash, February 26, 2022, https://www.youtube.com/watch?v=q5dy-NtVeFo&t=202s

EVIDENCE FOR COSMIC INFLATION

In an interview on the podcast *Closer To Truth* hosted by Robert Lawrence Kuhn, Alan Guth, the architect behind the theory of cosmic inflation, said,

> *"So, that theory became what we now call the conventional Big Bang theory: that the universe began in a hot, dense state and has been expanding ever since, and by understanding the physical processes involved—and there's not much to understand here, really. The model of the Big Bang really only incorporates the physics of simple gravity. One normally uses general relativity, but in fact, you can use Newtonian gravity, even, and you'll get pretty much the same answer. And then one just models an expanding gas and talks about how fast that would thin out and cool, and one calculates everything in terms of that, and it works very well. And it allows you to calculate the age of the universe in terms of present parameters like how fast the universe is expanding today. That now has been measured quite accurately, and especially when one combines that with information from the cosmic background radiation, one can now estimate the age of the universe back to that hot, dense initial state. It's said to be 13.7 billion years to an accuracy of .02 billion years, incredible accuracy. Now that still leaves the question of that hot, dense state was really the beginning, or whether or not there may be some prehistory to that. Everything I've said so far is really not controversial. I think all cosmologists agree that our local universe began from a very hot, dense state approximately 13.7 billion years ago. The possible prehistory is a more controversial thing, more speculative. The certainly could have been a prehistory, Various theories predict there probably was a prehistory. The theory I've worked on called inflation seems to imply there was almost certainly a*

> *prehistory, but still there would be a beginning someplace if inflation is correct."*[38]

By "prehistory," Guth means the theoretical period of time that elapsed between the origin of the universe and the Big Bang, however long it may have been. There is no reason to believe the prehistory was longer than cosmic inflation itself, which was only a tiny fraction of a second long. However, no matter how long the prehistory was, Guth seems clear that a true origin of the universe preceded the Big Bang and cosmic inflation.

Tufts University's Director of Cosmology, Alexander Vilenkin, one of the key proponents of inflation theory, said,

> *"Inflation actually is the theory of the bang, of the Big Bang. Inflation is the extremely fast, accelerated expansion of the universe. And it is driven by a peculiar form of matter, which is called the false vacuum. The most peculiar thing about it is that it is a very dense material. Its cubic centimeter has mass, at least the mass of the moon, and probably much higher. But the most peculiar property of this thing is that it has strong repulsive gravity. And this is what causes the universe to expand at a staggering rate. And since this force persists, the universe expands faster and faster. So, it gets huge in a tiny interval of time. So, this false vacuum eventually decays. It is unstable. And its energy turns into a hot fireball of elementary particles and radiation. From that point on, the gravity of particles is normal. It is attractive gravity. But this fireball expands by inertia because it was expanding so fast to begin with, and the end of inflation or the creation of the fireball is what we call the Big Bang. The expansion of the universe that we see today is just*

---

[38] Guth, Alan. "Alan Guth — Did Our Universe Have a Beginning?", Closer To Truth, April 11, 2021. https://www.youtube.com/watch?v=qv-g4JvIy6A

> *the consequence of the force of this initial explosion."*[39]

According to Vilenkin, cosmic inflation ends with the Big Bang. However, Arvin Ash has the opposite opinion about which came first.

Ash said,

> *"Inflation is believed to have occurred in our universe shortly after the Big Bang for an extremely short period of time, perhaps only 10^{-36} seconds. But in this short time, the universe expanded by a factor of 10^{78} in size. The evidence for inflation is the observation that our universe is homogeneous, that is, it looks about the same everywhere in space and appears to be geometrically flat. Why inflation ended within the short time that it did is not known. It's possible that on scales much larger than our observable universe, the universe is not so homogeneous. Since quantum mechanics ensures that there will always be some randomness, that inflation could last a bit longer or a bit shorter than expected in different parts of the universe. In the 1980s, Paul Steinhardt, Andrei Linde, and Alexander Vilenkin realized that the exponential expansion of cosmic inflation, even though it stopped in our part of the universe, could continue without end in other unobservable parts of the universe. And if that's the case, then the universe we are familiar with*

---

39 Vilenkin, Alex. "Mysteries of Cosmic Inflation — Episode 906", Closer To Truth, August 20, 2020.
https://www.youtube.com/watch?v=7IcooGU3jcU

may be a vanishingly small fraction of all that exists."[40]

Who is right? Does it even matter? The more important facts would seem to be the precision and duration of inflation, not whether it came immediately before or immediately after the Big Bang. Inflation happened so fast that if you blinked, you missed it. What difference does it make whether inflation ended with the Big Bang or immediately followed it? Katie Mack from the Perimeter Institute said:

> *"The Big Bang singularity could be in the universe's past, but not in any particular place, and never in a single point. As it turns out, though, even an infinitely large singularity doesn't work well with the cosmic microwave background. The CMB is just too perfect. The CMB light is incredibly uniform. If it really does come from the glow of the primordial plasma, that means the plasma was about the same temperature everywhere—to just one part in 100,000. But in the standard Big Bang model there's no reason that the primordial plasma billions of light years apart, on opposite sides of the sky, should ever have been the same temperature. Normally, for two things to come to the same temperature, they have to be in contact for a while."*[41]

---

[40] Ash, Arvin. "What Was There Before the Big Bang? 3 Good Hypotheses!", Arvin Ash, April 13, 2024, https://www.youtube.com/watch?v=QAuf97BhgRY

[41] Mack, Katie. "Cosmic Inflation Explained | Cosmology 101 Episode 6", Perimeter Institute for Theoretical Physics, August 9, 2024, https://www.youtube.com/watch?v=RHBlJHL7cBs

The more critical point is that inflation had an extraordinarily precise beginning and end. Alan Guth explained the precision of inflation, saying,

> *"This flatness [of the universe] problem is the statement that to get the universe to come out right, to look anything like what we see now, if you look back at what things must have been like at one second after the Big Bang, the expansion rate has to have been just right, or the mass density has to have been just right to an accuracy of one part in ten to the fourteen. Otherwise, the universe either rapidly re-collapses or expands so fast that no galaxies formed. This idea struck me very much at the time. I was bowled over by it. But I had no idea, really, what to do about it at the time. I just tucked it away in the back of my mind. Also in the fall of 1978, the other important event that got things rolling was a fellow post doc of mine at Cornell named Henry Tye came to me one day and asked me about magnetic monopoles and grand unified theories. He had said, well why don't we try to figure out how many of them would have been made in the Big Bang. At first this idea sounded crazy, and he really had to twist my arm to get me to work with him on it. When we did, we discovered that the universe would be swamped with magnetic monopoles if one combined the standard idea from grand unified theories with standard cosmology. So, that led us to start to scratch our head about whether or not there's any way to modify standard cosmology to make it consistent with grand unified theories and the observed absence of magnetic monopoles and the real universe. We came up with the idea that these magnetic monopoles would be extraordinarily suppressed if in the early universe, there was an extreme amount of super cooling, going below the temperature of the phase transition, without the phase transition yet happening. We assumed that this super cooling would have no effect on the*

> *expansion rate of the universe, and in our calculations, we just put in the standard expansion rate. And then the real excitement happened one night when I went home and looked at the equations that described how mass densities effect the expansion of the universe, which are all really standard equations of cosmology. It became immediately apparent that the consequences of this super cooling would result in a form of matter that would really turn gravity on its head and cause it to become repulsive instead of attractive and that that would propel the expansion of the universe, and perhaps even explain why the universe is expanding, which is what became the theory of inflation itself. And I realized that same night, it was extraordinarily exciting, that this process of exponential expansion would actually solve this flatness problem. The expansion rate that is driven by this gravitational repulsion is just the right rate to drive the universe to this critical density that the universe needed to have, to extraordinary accuracy, in the early period."*[42]

The best evidence for inflation is cosmic microwave background radiation, the afterglow of the creation of the universe from nothing, either by God or by serendipity. The CMB shows a uniformity in the distribution of heat throughout the universe, which matches the predictions of inflation theory. The idea is that the scientific evidence doesn't match the Big Bang alone; it almost perfectly matches the Big Bang plus cosmic inflation.

To be brutally honest, I don't claim to fully understand cosmic inflation. (Nor do I claim to fully understand anything else, for that matter.) I mention the scientists below because they are the acknowledged authorities on cosmic inflation. Those with

---

[42] Ash, Arvin. "Mysteries of Cosmic Inflation — Episode 906", Closer To Truth, August 20, 2020. https://www.youtube.com/watch?v=7Ic0oGU3jcU

interests beyond a superficial understanding of the subject should read books written by these authority figures on the subject, assuming these books exist. At a minimum, there should be scientific papers that describe their work in greater detail.

Standard Inflation Model

Of the physicists associated with cosmic inflation, Alan Guth is perhaps the most famous for his contributions. A professor at M.I.T., Guth is considered by many to be the primary architect of cosmic inflation theory. He originally proposed the idea in 1979 after contemplating the lack of magnetic monopoles (a hypothetical particle that is a magnet with only one magnetic pole) in our current universe. Guth was one of the three scientists who developed the Borde-Guth-Vilenkin theorem, which says any expanding universe must have a finite past. Another way of saying this is to say the currently expanding universe means our universe once had an origin.

Together with Andrei Linde and Paul Steinhardt, Guth shared the 2002 Divac medal for developing the theory of cosmic inflation. Together with Alexei Starobinsky and Andrei Linde, Guth received the 2014 Kavli Prize in Astrophysics for his pioneering work on the theory of cosmic inflation. If one scientist should be declared the father of cosmic inflation theory, it would have to be either Alan Guth or Alexei Starobinsky.

Initial Inflation Model

Alexei Starobinsky developed the first model of cosmic inflation (known as Starobinsky inflation) by modifying Einstein's theory of general relativity. However, his work was largely unknown outside of the Soviet Union, and Alan Guth became somewhat more famous for having the same general idea. This is not unlike the situation involving Charles Darwin and Alfred Russel Wallace, both independently developing the idea of natural selection at roughly the same time. Starobinsky was a brilliant physicist who, together with Yakov Zeldovich, famously proposed black holes can emit energetic particles (while Starobinsky was still a student), leading to the discovery of Hawking radiation as a quantum effect near the event horizon of a black hole.

Together with Alan Guth and Andrei Linde, Starobinsky shared the 2014 Kavli Prize in Astrophysics for his work on cosmic inflation. Ironically, both men reached the same basic conclusion independently, which added credibility to the idea they were onto a special idea that enhanced the explanation of the Big Bang and more fully explained the origin of the universe.

Bouncing Universe Hypothesis

Physicist Paul Steinhardt was an early advocate of cosmic inflation, working with Alan Guth and Andrei Linde to develop the theory. Together with Andrei Linde and Alan Guth, Steinhardt shared the 2002 Dirac medal for developing the theory of cosmic inflation. More recently, Steinhardt has become a vocal critic of inflation, claiming that data from the Planck satellite contradicted the theory because they expected to find evidence of cosmic gravitational waves but didn't find that evidence.

As an alternative, Steinhardt advocates for a "big bounce" or bouncing universe cosmology, where a period of contraction for a previous universe preceded the expansion of this current universe. The bouncing universe hypothesis avoids the controversy of the origin of matter by assuming matter has always existed, but it doesn't explain the current configuration of that matter into this universe.

Eternal Chaotic Inflation Model

Andrei Linde initially took Guth's inflation theory and modified it, calling it new inflation. It differed from Guth's model in that inflation continued while the scalar field rolled down. Linde has since developed a new form of inflation called eternal chaotic inflation, in which inflation is part of the eternal multiverse, and occasionally a new universe forms like a bubble.

With Alan Guth and Paul Steinhardt, Linde shared the 2002 Divac medal for developing the theory of cosmic inflation. Together with Alan Guth and Alexei Starobinsky, Linde shared the 2014 Kavli Prize in Astrophysics for his work on cosmic inflation. What do all of these physicists have in common? They all believe "our" universe briefly expanded at an exponential rate from a distant point in time. They also believe their personal interpretations of this period of expansion are the best or most correct interpretations. Everybody has an ego, and everybody thinks they can provide the best, most comprehensive solution to the problem.

The problem with the hypothetical multiverse is the same as the problem of looking for particles that might only exist in a hypothetical sense—we're deliberately contaminating the information we know with assumptions we will never prove. The thing is, without cosmic inflation and the Big Bang, we don't have this universe. The universe did not exist, and then that tiny dot of material that would become the universe suddenly began to exist. The tiny dot of material might have been the debris of a previous universe, but that doesn't really matter when it comes to the existence of the current universe. That tiny dot of material rapidly expanded to become this universe. That's the important thought.

ANALYSIS OF COSMIC INFLATION

However, if what all these experts say is true and cosmic inflation occurred in some form, the precision timing of that fraction of a second is both unbelievable and uncanny. According to the experts, if the universe expanded at a rate even one tiny fraction of a second longer, the universe would allegedly continue expanding so fast it couldn't form stars or planets. Likewise, if the universe's expansion were even a minuscule fraction of a second shorter, the universe allegedly would collapse back on itself. The timing of the expansion was exquisitely perfect in duration. It was another Goldilocks enigma—the cosmic expansion was neither too long nor too short. It was just right.

Katie Mack asks an interesting question:

> *"So, how did two points outside each other's observable universe agree to be exactly the same temperature? This is called the horizon problem because different parts of the sky seem to have somehow communicated despite being outside each other's cosmic horizons. The most widely accepted solution to this is to insert a period of extremely rapid expansion called cosmic inflation in the first tiny fraction of a second of the cosmos. Inflation fits between whatever the beginning was, possibly a singularity, and the hot Big Bang, when the universe was hot and dense. During inflation, space was expanding quickly enough to take a tiny, tiny region of space that in those first moments had already become uniform and blow it up by more than 26 orders of magnitude, making it larger than our entire observable universe all within a billionth of a trillionth of a trillionth of a second. Then the energy field causing inflation decayed, filling the cosmos with radiation, and the expansion proceeded at a more orderly pace." Later in the video, she admits, "We still don't have a solid theory for what started inflation or why it stopped."*[43]

---

43 Mack, Katie. "Cosmic Inflation Explained | Cosmology 101 Episode 6", Perimeter Institute for Theoretical Physics, August 9, 2024,

Speak for yourself. I have a fairly solid theory about cosmic inflation.

Planned Versus Unplanned

The extraordinary precision of cosmic inflation makes it extraordinarily difficult to believe it might have been unplanned. The mathematics simply do not favor the unplanned version of inflation, where an intelligent control is neither responsible for inflation's beginning nor end. It is even more remarkable, considering that inflation occurred at such a precise time relative to the Big Bang, yet has its own separate and specific calculations of improbability associated with it. It's almost (or exactly) as if the two events were perfectly coordinated. Arvin Ash efficiently summarizes the status of cosmic inflation theory from one perspective, saying:

> *"So, our most accepted current standard theory of the Big Bang goes something like this: about 13.8 billion years ago, the Big Bang happened. We don't know what caused it or what happened before this event. But about $10^{\wedge-32}$ seconds later, cosmic inflation happened, causing the universe to expand exponentially. Then it ended. And then the universe continued to expand, but at a much slower pace. And these events largely explain what we observe in the universe today, 13.8 billion years later. With eternal inflation, the story changes to something like this: we don't know what happened at the beginning. But a fraction of a second later, inflation happened. This could have happened arbitrarily far in the past. It is also possible that there may not have been a beginning at all. Many bubble universes formed and are continuing to form. At some point within this space, our bubble formed. It was probably not the first and is certainly not the last.*

---

https://www.youtube.com/watch?v=RHBlJHL7cBs

> *Inflation will continue forever continually sprouting new bubbles of universes in the space that is created within it. But what we observe in our bubble 13.8 billion years after its formation can be explained by the Big Bang model. So, what we think of as the start of the universe may just be the point at which inflation ended in our part of the universe."*[44]

Not to quibble, but the order of events for the Big Bang and cosmic inflation it isn't quite as certain as Ash implies. There isn't unanimous agreement. Either as Ash said, the Big Bang happened and then cosmic inflation immediately followed, or cosmic inflation ended with a Big Bang. It sort of depends upon whom you ask.

Stephen Hawking was neither a fan of the multiverse nor eternal inflation: "The usual theory of eternal inflation predicts that globally our universe is like an infinite fractal, with a mosaic of different pocket universes, separated by an inflating ocean," he said in an interview. "The local laws of physics and chemistry can differ from one pocket universe to another, which together would form a multiverse. But I have never been a fan of the multiverse. If the scale of different universes in the multiverse is large or infinite, the theory can't be tested."[45]

The question isn't whether or not one should be a fan of multiverse hypotheses. The question is whether one can explain the existence of this precise, fine-tuned universe without either

---

[44] Ash, Arvin. "Eternal Inflation: The BEST MULTIVERSE Theory of Reality", Arvin Ash, March 11, 2022, https://www.youtube.com/watch?v=nziePav5OMg&t=113s

[45] Hawking, Stephen and Hertog, Thomas. *"A Smooth Exit from Eternal Inflation?" Journal of High-Energy Physics (2018). DOI: 10.1007/JHEP04(2018)147",* https://www.cam.ac.uk/research/news/taming-the-multiverse-stephen-hawkings-final-theory-about-the-big-bang

appealing to a creator God or invoking a multiverse hypothesis. In his book *A Brief History of Time*, Hawking famously wrote, “If the rate of expansion one second after the big bang had been smaller by even one part in a hundred thousand million million, the universe would have re-collapsed before it ever reached its present size.”[46] I don’t know about you, but one in a hundred thousand million million sounds like an absurdly precise value. Skeptics and other critics have complained that creationists like to ignore that Hawking also wrote a few pages later, “Moreover, the rate of expansion of the universe would automatically become very close to the critical rate determined by the energy density of the universe. This could then explain why the rate of expansion is still so close to the critical rate, without having to assume that the initial rate of expansion of the universe was very carefully chosen.”[47]

However, the first statement appears to be a statement of fact, and the second statement sounds like Hawking’s opinion that inflation might solve the fine-tuning problem if we can exercise enough wishful thinking.

CONCLUSION

What do these brilliant scientists have in common? These men and women all study the structure and origin of the universe. They all agree that a brief period of exponential expansion helped the universe become what it is today—a place capable of supporting complex living organisms here on Earth. The vast majority of these scientists are probably non-religious, because sadly, in the pride of their own intellect, they seek to explain our most complex problems without the need to invoke a deity. They all understand the universe had to be stretched for their

---

[46] Hawking, Stephen. *A Brief History of Time*. Page 126. New York. Bantam. 1996. Print.

[47] Ibid. Page. 133.

calculations to make sense, and each looks to put his own unique spin on the solution to this most mysterious problem. Some modern atheists like to suggest the Bible was written by a group of illiterate shepherds and carpenters, but the reality is that the people who wrote the Bible display a reasonably accurate and quite sophisticated knowledge of early universe cosmology. Also, illiterate people don't know how to read or write. That's what illiterate means. Interestingly, the Bible describes cosmic inflation.

Isaiah 42:5 says:

> *"Thus says God the LORD, who* ***created the heavens and stretched them*** *out, who spread forth the earth and that which comes from it, who gives breath to the people on it, and spirit to those who walk on it."*

Jeremiah 10:12 reads,

> *"He has made the earth by His power, He has established the world by His wisdom and has* ***stretched out the heavens*** *at His discretion."*

Both of those verses describe God stretching the heavens in the past tense. Those words were not carelessly chosen.

The Bible also describes God as continuing to expand the universe.

Job 9:8 says:

> *"He [God] alone* ***stretches*** *out the heavens and treads on the waves of the sea."*

The book of Isaiah reinforces the point in multiple chapters.

Isaiah 40:22 says,

> *"He sits enthroned above the circle of the earth, and its people are like grasshoppers. He **stretches** out the heavens like a canopy and spreads them out like a tent to live in."*

Isaiah 44:24 adds,

> *"Thus says the LORD, your Redeemer, And He who formed you from the womb: 'I am the LORD, who makes all things, **Who stretches out the heavens** all alone, Who spreads abroad the earth by Myself;'"*

Finally, Zechariah 12:1 reads:

> *"The burden of the word of the LORD against Israel. Thus says the LORD, **who stretches out the heavens**, lays the foundation of the earth, and forms the spirit of man within him."*

It turns out that the scientific theory of cosmic inflation is very well-represented in multiple verses and multiple books in the Bible.

MIRACLE 3
THE ORIGIN OF LIFE

INTRODUCTION

Amino acids are the building blocks of peptides. Peptides are short chains of amino acids connected by peptide bonds. Longer chains of peptides are called polypeptides, and once polypeptides achieve a certain mass, they are called proteins. Proteins (plus DNA) are considered the building blocks of cells. The smallest form of life is an individual cell, or more precisely, a single-celled organism. The transition from inanimate matter to a living organism is a mystery that secular minds have been trying to solve through a series of small, incremental steps. The first step in the process is the synthesis of amino acids, which scientists in the 1950s demonstrated in the famous Miller-Urey experiment. However, progress since that experiment has been few and far between. That hasn't stopped origin-of-life researchers from making some audacious claims, but those claims have typically failed to withstand closer scrutiny.

There are two basic ideas regarding how life came to exist on Earth. One concept (embraced by atheists, materialists, and naturalists) is that life began as a single cell, in its simplest form at the very base of the tree of life. That first organism slowly built its way up to becoming more complex over a very long time. There are numerous problems with this idea, but it must exist because it serves as one of two basic alternatives. The other basic idea is that life was created by some intelligent entity that some people like to call "God," and if we don't allow for any alternative, we are implying that we are certain God exists, and that is as problematic as insisting that evolution is the *only* possible explanation for the diversity of life.

The idea that works from the bottom-up is by a collaboration of a whole series of unplanned, undirected processes—the fine-tuned Big Bang and cosmic inflation produce the universe, and then what happens purely by accident to be the perfect elixir of building blocks, the first simple(r) cell forms, and once started, life is off to the races. The other idea is that the source of the intelligence that created our universe and stretched it to become habitable for life, then proceeded to create living organisms. Of course, there are variations such as theistic evolution, where the source of intelligence created the universe and the first life form but then got lazy, but variations may be disregarded because the two basic ideas (planned or unplanned) cover all the bases.

In truth, theistic evolution is an oxymoron because evolution, by definition, is an unplanned process, and theism involves a planner. The only reason people believe that unplanned abiogenesis might even be possible is because we know life exists, and some people simply don't like the idea of involving a planner. We all suffer from some form of cognitive dissonance, in which we cannot help but hold a bias toward an outcome we prefer. We cannot say we know with absolute certainty that the universe and life are planned, but we *can* safely say that a planned universe is more logical and makes the most sense. In an unplanned universe, there is no reason to assume the cosmic variables that resulted in a fine-tuned Big Bang had a choice except to have the crucial values that result in success. There is every reason to believe the values could have been different, and even the slightest change spells disaster for the universe we occupy.

With a planned universe, we don't need multiverse hypotheses. The problems of cosmic inflation may be considered part of the overall design. We don't need a cyclic or bouncing universe to explain away the origin of matter. We only need to allow for the possibility of the planner with a terrific plan.

The most popular current belief about the origin of life is that life initially formed as single-celled organisms called extremophiles near nutrient-rich thermal vents in the ocean. These spontaneously formed extremophiles then consume chemicals or absorb light (photosynthesis) for nutrition. They reproduce asexually. From these very simple forms of life, every organism on modern Earth has descended, with lots of modifications over very long periods of time.

However, if unplanned abiogenesis is true, then an unplanned Big Bang and unplanned cosmic inflation must also be true, and the improbability calculations associated with each event must be taken into consideration accordingly.

The probability of a fine-tuned Big Bang is said to be extremely low; some estimates put it at 1 in 10^300. The probability of cosmic inflation's precision was estimated by Stephen Hawking as being refined on the order of one in a thousand million-million. Fred Hoyle compared the origin of life problem to a tornado flying through a junkyard and producing a new jet airplane. Each of these numbers represents an absurdly small fraction of a single percent. If the goal was to try to calculate the probability of our unplanned and undirected existence, we would need to multiply these improbabilities together to produce a cumulative estimate. These are compounded improbabilities.

Life itself is the evidence that life must have had an origin, just as THIS universe apparently had an origin. While evolution is believed to be the best explanation for the diversity of living organisms, it is critically important to note that life cannot evolve until it exists. Before evolution can ever become possible, creation (either created by God, luck, or due to some "natural" law—nothing isn't really an option) has already occurred.

Biologists love to claim that evolution is an indisputable fact, but the fact is that the supposedly unchallengeable theory relies entirely on the success of an unproven hypothesis.

For a moment, let's imagine that the atheistic worldview of Susan Blackmore, Sam Harris, and Richard Dawkins is true: our sense of individuality is a delusion. Our thoughts are merely the result of chemical reactions in our brains. Our DNA and our environment determine our actions. In this bleak materialist worldview, life cannot be said to have any true purpose or meaning.

EVIDENCE FOR THE ORIGIN OF LIFE

Advocates of Darwinian evolution like to point out that the diversification of life is a separate biological process. They say abiogenesis is a chemical problem that has no impact on the ironclad nature of the evidence for evolution. My response has consistently been to point out that life cannot evolve until it exists. Any so-called evidence for evolution, no matter how strong it is perceived to be, still suffers from the fatal weakness of relying entirely on the success of an unproved and virtually impossible hypothesis before it can ever become feasible.

The Miller-Urey Experiment

Stanley Miller and Harold Urey performed an experiment in the late 1950s in which they combined methane, water, hydrogen, and ammonia, intended to mimic the expected atmosphere of early Earth, inside a series of sealed, sterile glass tubes and flasks connected in a loop so the mixture could circulate. The experiment demonstrated that some amino acids could spontaneously form from a mixture of pre-organic materials under certain conditions, but amino acids are not remotely comparable to a living organism. The amino acids produced by the experiment did not have the correct chirality that would facilitate the origin of life, but it did result in the creation of amino acids, which was a significant accomplishment. The

Miller-Urey experiment attempted to simulate a "pre-Earth" atmosphere in a laboratory. Raw chemicals were charged with electricity to mimic lightning strikes to provide a catalyst for chemical reactions to occur, and a few amino acids were created as a result. Scientists have lauded the results of the experiment as evidence that life could have randomly formed by chance, but there is a vast difference between an amino acid and a nucleotide, and even bigger differences between an amino acid and DNA.

Miller-Urey was certainly a very interesting experiment and a remarkable achievement, but its importance is overstated. Saying that Miller-Urey gives us crucial insight into the origin of life is kind of like claiming we know how to build the Empire State Building because we know how bricks can be made. Amino acids typically have a chirality (meaning they are either left-handed or right-handed), and the amino acids produced by Miller-Urey had the wrong chirality or no chirality. These amino acids would be useless for protein formation. The chemicals used in the experiment were chemicals known to be present in amino acids. There was no contamination of non-essential material.

The entire experiment proved to be an exercise in intelligent design. The chemists combined the right mixture to work toward a specific goal, an intended result. To truly replicate an unplanned event, a wide assortment of chemicals should have been thrown into the mix, and then hope for the best.

Avin Ash clearly understands the problem:

> *"Abiogenesis is not evolution. Evolution is the process of development and diversification of living things from earlier living things. Evolution does not say anything about how life first originated. So, how did life originate? Despite the incredible variations that we see today at the fundamental level, all living things contain a trinity of elements. First, nucleic acids,*

> *which make up the DNA or its simpler form, called RNA. These contain the blueprints of life and are self-replicating molecules. Second, there are proteins, which are the workhorses that perform the important functions of your body. And third, there are lipids, which encapsulate the cells of your body. Before any living things existed, before animals, plants, and even bacteria existed, these three things had to have been present in the primordial soup in order for life to start. Some argue that the most important component of this trinity are the lipids, which make up the cell walls. Why would they be most important? Because without the wall, or a way to encapsulate certain elements within the soup, there would just be a soup of material that would just be disorderly and floating around in a sea of liquid. It would not be functioning inside something that could potentially self-replicate. But because these lipid membranes could potentially form around other elements, they could bring disparate parts of various chemicals together that could potentially interact, combine, and work together to perhaps eventually form a machinery for self-replication."*[48]

That's a lot of wishful thinking, though, isn't it? Conditional words like "could", "perhaps", "might", and "maybe" are being asked to do a lot of heavy lifting. The unplanned Big Bang has one improbability value. Cosmic inflation has its own improbability value; abiogenesis is a third improbability to consider as part of the unplanned universe. To say there is a zero percent probability that the unplanned universe is true implies an imperfect understanding of statistics and probability. The hope for the success of the unplanned universe

---

[48] Ash, Arvin. "How did life begin? Abiogenesis. Origin of life from nonliving matter", September 6, 2019. https://www.youtube.com/watch?v=nNK3u8uVG70&t=600s

starts out extraordinarily low with the Big Bang and only gets worse as we move on to other necessary miracles.

Ash said,

> *"You might say, 'Well, that's fine and dandy, but having all the precursors get together inside a lipid cell wall does not necessarily mean they will all come together to form a self-replicating living cell. How do the complex molecules come together to self-replicate and become a living organism?' And if I'm being honest, this is currently not well understood, and there's no experiment or smoking gun evidence right now that points to a precise mechanism of how this could have happened."*[49]

I do appreciate the honesty. Arvin Ash's videos are very informative, and he's got a great personality for this sort of contribution to expanding the knowledge base of humankind. He's a powerful advocate for science and very engaging. There is no smugness, no condescension, no air of superiority in his presentation. Ash communicates in a very conversational tone. Even when he says something I might not agree with, he doesn't say it with an attitude that he can't possibly be wrong, and you'll be a liar if you even dare to challenge him. Contrast Ash's style with "Professor Dave" Farina and you'll have two completely different experiences. One will make you feel smarter for having watched his video, and the other will make you feel foolish for having wasted your time. Yet Farina has more than three times as many followers as Ash, which seems to indicate people prefer sophomoric buffoonery to real science information. Almost four million people have been attracted by videos with titles such as "Grifter Tier List: Who is the biggest scumbag I've debunked?",

---

[49] Ash, Arvin. "How did life begin? Abiogenesis. Origin of life from nonliving matter", September 6, 2019.
https://www.youtube.com/watch?v=nNK3u8uVG7o&t=600s

or "Avi Loeb is a Fraud Now", or "Terrance Howard vs. Russell Brand: Ultimate Douche Warriors."

Personally, I do not find that sort of material interesting. It's too overtly hostile for my taste. But I digress. Ash concedes,

> *"There are creationist arguments such as the one that says if you put all the parts of a watch into a big vat and keep stirring it for a million years, a functioning, ticking watch is not going to magically form inside the vat. Some cite an estimate by scientists Fred Hoyle and Chandra Wickramasinghe showing that the probability of all the chemicals in a simple bacterium arising on their own by chance is something like 1 in* $10^{40,000}$ *power which is more than the Planck volume of the entire universe. So, that is a virtual impossibility. But this number and the clock parts in a vat argument are oversimplifications. They ignore the fact that sophisticated life forms like current day bacteria almost certainly did not arise spontaneously but arose in much simpler incremental steps that had a much higher chance of occurring."*[50]

For the sake of argument, let's say abiogenesis is truly possible, and that explains how the first organism formed on Earth. How did we take the next step? That organism had DNA or, at a minimum RNA, correct? It reproduced asexually, meaning it cloned itself, producing exact copies of itself. From where does any new DNA information come to form an evolved descendant—a virus? Then what created the virus?

Please note that earlier, when I wrote about the Big Bang and used the phrase "virtual impossibility" to describe the odds of success for an unplanned event, I took the extra step of

---

50 Ash, Arvin. "How did life begin? Abiogenesis. Origin of life from nonliving matter", September 6, 2019.
https://www.youtube.com/watch?v=nNK3u8uVG70&t=600s

explaining what I meant by virtual impossibility, which is an improbability so grotesque and absurd that it barely merits serious consideration. A true impossibility has a zero percent chance of success. That represents absolute certainty or provable knowledge, which only a fool would claim to possess. I might be foolish, but I'm not *that* foolish. To say that there is zero percent chance of success for the origin of the universe (or the origin of life) is essentially a faith-claim. Claiming there is absolutely no chance (truly impossible) that the unplanned universe scenario with an unplanned origin of life could occur is basically a knowledge claim. I would essentially be claiming to **know** God created the universe, not that I merely have a **very strong belief** God created the universe. How could I claim to know what I know I cannot prove? That would make me either a liar or a fool. I prefer to think of myself as neither.

I once listened to a lecture by Stephen Hawking, and he said that we cannot know with absolute certainty that the entire universe wasn't fabricated yesterday and we've only existed for one day, with a lifetime's worth of false memories created by some form of higher intelligence and implanted in our organic brain. Now, that idea sounds preposterous to me. I have no reason to believe it is true...but can I prove it isn't true? How could I ever prove that the claim is false? It seems safe to assume that the claim is false because it is unnecessarily complicated and thus the most complex solution to the problem. As H. L. Mencken famously said, every human problem has a well-known solution that is neat, plausible, and wrong.

This reminds me of my favorite line from the movie Red Dragon, spoken by the character Hannibal Lecter: "Our scars have the power to remind us that the past is real."

As I type these words, my left thumb subconsciously rubs over the scar on my left forefinger from when my sister accidentally cut it with a pair of scissors when we were young and playing together. If I were to call my sister right now, there is little doubt in my mind she would confirm my memory of the accident is quite real, which means my programmer would have also had to know to program that same memory into my sister. There is a slight bump at the bottom of the scar because one of my stitches burst, and we didn't bother getting my finger restitched because it had almost healed. Indeed, if my past is merely a programmed figment of my imagination, the programmer paid some truly extraordinary attention to detail.

Ash offers the standard rebuttal argument to the improbability of a cell by saying the cell assembled in a series of smaller steps, just as life began as a single cell and became more complex over time. In this unplanned and undirected scenario, Time is the magic elixir to solve virtually any problem. The problem with Time is that it eliminates observation from the scientific method, and without observation, we are literally blind. You might guess the right answer, but without seeing for yourself, you'll never know for sure. Science has been described as like a detective coming to the scene of the crime after the fact and piecing together the evidence to solve the mystery, but evidence collected after the fact is considered circumstantial. Clues (even DNA) can be planted, and fingerprints don't come with a timestamp saying when they were left. The best (direct) evidence is always going to be eyewitness testimony from credible witnesses. Ash explains the argument for an unplanned origin of life, saying,

> *"There are stats such as the one that says the odds of creating a protein molecule by chance are 1 in 10* ^{45}*. Odds such as these and others not only ignore the idea of simpler precursors but also ignore the fact that it was not just one set of amino acids at one place and one time, but it was trillions and trillions of amino*

> *acids reacting in countless places over millions of years that resulted in simple protein molecules."*[51]

Entropy is the problem with that argument. Time doesn't tend to create; Time tends to destroy through decay and erosion. There aren't millions of years available for these unplanned processes to work because Time is working to destroy these molecules faster than they could form. The shelf life of unprotected DNA is quite short.

A Nobel Prize for Attempting to Create a "Protocell"

Jack Szostak is a Nobel Laureate, a Professor of Genetics at Harvard University, and an origin of life researcher who has tried to replicate the chemistry on Earth before life began. Szostak has been trying to create a simple "protocell" in his laboratory that could theoretically evolve into a more modern cell if given enough time.

An article written about Szostak's work at Harvard says,

> *"Life was probably born in a small pond or lake, Szostak believes, not in an ocean, as many people think. Rain-fed pools provide a freshwater environment, compatible with the delicate cell membranes formed from simple fatty acids, which would be destroyed instantly in the salty oceans. Some such pond was the place where crucial elements were mixed, heated, and cooled in the right sequence to become life. Inanimate molecules, congregated together inside a fatty skin, somehow became capable*

---

[51] Ash, Arvin. "How did life begin? Abiogenesis. Origin of life from nonliving matter", September 6, 2019. https://www.youtube.com/watch?v=nNK3u8uVG7o&t=600s

> *of replication, and of evolution: the definition of life."*[52]

In that same article, Szostak also said,

> *"We're looking at a very narrow slice of the whole problem. We assume we have the chemical building blocks of life: the question we're looking at is, what do we need to do to make these chemicals get together and work like a cell?"*[53]

That's one heck of an assumption, isn't it? That the chemical building blocks of life will conveniently be there right when you need them? I am reminded of the joke about the scientists who challenge God to a contest to see who could make the best human from clay. When the scientist's turn comes, they bend down to scoop up some clay, but God says, "Hey! Get your own dirt!"

When Szostak was awarded a share of the Nobel Prize, the following paragraph was written by the Nobel organization describing his work that earned the special recognition:

> *"An organism's genes are stored within DNA molecules, which are found in chromosomes inside its cells' nuclei. When a cell divides, it is important that its chromosomes are copied in full, and that they are not damaged. At each end of a chromosome lies a cap or telomere, as it is known, which protects it. After Elizabeth Blackburn discovered that telomeres have a*

---

52 Unknown. "Making Life from Scratch: Biochemist Jack Szostak's search for the First Cell", (2012). Harvard University. https://origins.harvard.edu/pages/research-spotlight-jack-szostak

53 Unknown. "Making Life from Scratch: Biochemist Jack Szostak's search for the First Cell", (2012). Harvard University. https://origins.harvard.edu/pages/research-spotlight-jack-szostak

> *particular DNA, through experiments conducted on ciliates and yeast, she and Jack Szostak proved in 1982 that the telomeres' DNA prevents chromosomes from being broken down."*[54]

Working with David Bartel, Szostak conducted an experiment where they synthesized trillions of random RNA molecules and performed ten rounds of tests where molecules that could catalyze reactions that caused linkage to occur were selected for additional rounds of the tests, and the result after ten rounds was that it could be claimed the RNA had evolved. But given the fact that Bartel and Szostak specifically selected for molecules with an increased tendency to have a catalytic reaction, wouldn't that make the experimental results more accurately described as the product of intelligent design?

In an article published in *Scientific American* Szostak wrote, "It is virtually impossible to imagine how a cell's machines, which are mostly protein-based catalysts called enzymes, could have formed spontaneously as life first arose from nonliving matter around 3.7 billion years ago."[55] However, more recently, Szostak has changed his tune quite a bit and now claims that the origin of life is not "as hard as it looks."

Oh, really? If that's the case, why hasn't anyone created life in the laboratory yet? Perhaps abiogenesis hasn't been simulated because a living cell requires DNA, enzymes, and lipids all to exist before the cell can form. We don't know how one of these primary components of a cell forms, much less all three. It isn't

---

[54] Jack W. Szostak – Facts. NobelPrize.org. Nobel Prize Outreach 2025. Thu. 31 Jul 2025.
https://www.nobelprize.org/prizes/medicine/2009/szostak/facts/

[55] Richardo, Alonzo and Szostak, Jack. "The Origin of Life on Earth: Fresh clues hint at how the first living organisms arose from inanimate matter", Scientific American, September 1, 2009,
https://www.scientificamerican.com/article/origin-of-life-on-earth/

as if DNA could self-organize and then wait a million years for lipids to figure out how to form a cell membrane. The membrane needs to exist to protect the DNA from the external environment. DNA doesn't survive very long outside the confines of a cell. Enzymes are a special form of protein needed to perform work inside the cell, including during mitosis.

Almost Creating Life in the Lab

Ricki Lewis audaciously claimed that Craig Venter replicated abiogenesis in an article titled "How Craig Venter Created Life." I don't wish to belittle or minimize the accomplishments of Venter, but he most certainly has **not** created a living organism. What Venter did was take an existing cell and strip out the DNA inside before synthetically arranging the nucleotides of a known bacterium, and inserted that sequence into the stripped-out cell membrane. That synthetic cell began to reproduce. An extremely impressive accomplishment, to be sure, but calling what Venter did "the creation of life" is a rather grotesque exaggeration, given that he used an existing cell membrane and repeated a known DNA sequence.

The *Time* magazine article headline was a bit more honest: "Scientist Creates Life Almost." While still not entirely accurate, at least the use of the word "almost" makes it clear that Venter has not actually created life. It leaves open for debate how close "almost" comes to success. The truth is, not that close. To be fair, Venter isn't making these claims himself. Others have made these claims on his behalf. But it's also not like he's been protesting the public impression created by others on his behalf.

We know that cell membranes are composed of lipids, which means lipids are essential for the existence of cells. In other words, cells could not exist if lipids did not exist. Yet lipids are organic compounds, so how could lipids have existed before cells or organic chemistry existed? Venter skipped that part of the cell creation process with his synthetic cell. He didn't have to manufacture his own cell membranes or any enzymes because they were already part of the existing cell.

Please don't misunderstand—Venter's accomplishments are nothing short of remarkable, but nevertheless, they fall considerably short of a human being creating a living organism. It was more like creating Frankenstein than a new, living organism.

An Honest Confession

Chemistry professor and origin of life researcher Lee Cronin might be the most honest scientist involved in origin of life research after he posted a comment on the internet claiming origin of life research was a scam because nobody really believes it is possible. In a rare but refreshing moment of candor, Lee Cronin wrote on social media, "Origin of life research is a scam." When asked why he'd made that claim, Cronin said, "Because no one is really trying to answer the question or thinks it can be done."[56]

Is he right? Is the origin of life research a scam? Since that exchange, Cronin has subsequently claimed that his comment was meant to be taken tongue-in-cheek. He was only joking, he now claims. However, that new claim came a year and a half after he made the initial comment. He didn't explain the joke, if it was meant to be a joke.

When Dr. Cronin initially said origin of life research was a scam, he did have an explanation for what he meant to say, which was to say that nobody really thinks it can be done. Dr. Cronin also publicly admits he is an atheist. He could be biased toward the belief that a blind process subject to natural laws and random

---

[56] Cronin, Lee. "Atheist Scientist says Abiogenesis is not possible", YhwhScience, May 23, 2023, https://www.youtube.com/shorts/CeGU2OyQMVE

chance could produce RNA, which then might have evolved into DNA. Now proposed by Lee Cronin, assembly theory holds that we can look at any object in the universe and quantify how complex it is by finding out the number of steps it took to create the object. The object can be determined to have been built by a process like evolution by counting the number of steps to create the object and the number of copies that exist.

In the same interview on Lex Fridman's podcast where he claimed the "origin-of-life is a scam" comment was meant tongue-in-cheek, Dr. Cronin spilled the beans regarding Venter's claim of having created life in the lab:

> *"Coming back to the scam, the scam is if we just make this RNA we've got this fluke event we know how that's simple. Let's make this phosphodiester, or let's make ATP or ADP, now we've got this part nailed. Let's now make this other molecule, and another molecule. How many molecules are going to be enough? And then, the reason I say this when you go back to Craig Venter, when he invented his life form, Cynthia, this plat microbe or minimum plasmid...it's a myoplasma or something, I don't know the name of it, but he made this wonderful cell and said, 'I've invented life.' Not quite. He facsimiled the genome from this entity and made it in the lab, or the DNA, but he didn't make the cell. He had to take an existing cell that has a causal chain going all the way back to LUCA and he showed when he took out the gene, the genes and put in his genes, synthesized, the cell could boot up. But it's remarkable that he could not make a cell from scratch and even now today synthetic biologists cannot make a cell from scratch because there's some contingent*

information embodied outside the genome in the cell and that is just incredible."[57]

I like Cronin. Even though he's an atheist, he isn't an insufferable jerk about expressing his opinions. While he's clearly over-optimistic when it comes to the potential success for his own research, Cronin is remarkably honest about the current state of origin-of-life research.

The Man Who Invented the Gene Gun

John Sanford is an author, inventor, and former Cornell University professor. Dr. Sanford is probably best known for inventing the Biolistic Particle Delivery System, more popularly known as the "gene gun," but he also wrote a fantastic book titled *Genetic Entropy and the Mystery of the Genome* and has also published more than one hundred articles in scientific journals and publications. Ever heard the expression "genetically modified" food? That is food that has been genetically improved to produce a higher yield or to be more drought resistant. That feat is accomplished using Dr. Sanford's device, the gene gun. The so-called gene gun delivers external DNA, RNA, or proteins into a cell—it works best on plant cells. The word "biolistic" in the official name of the device is formed by combining the words biology and ballistic because the device uses high pressure gas to "shoot" the genetic material into the targeted cell.

According to Dr. Sanford, DNA is incredibly complex, having different meanings on multiple levels despite multiple constraints on how the information in the strand is interpreted. DNA is an elegant and sophisticated code. It was once believed that large sections of DNA were ancient remnants of DNA that

---

57 Cronin, Lee. "Lee Cronin: Origin of Life, Aliens, Complexity, and Consciousness | Lex Fridman Podcast #269", Lex Fridman, March 11, 2022, https://www.youtube.com/watch?v=ZecQ64l-gKM

served some purpose in ancestral creatures, but were known as Ancient Repetitive Elements (AREs), also known as junk DNA. Any DNA that didn't code for protein was assumed to be junk DNA—except in 2024, Victor Ambros and Gary Ruvkun were awarded the Nobel Prize in Physiology or Medicine for their discovery of microRNA, molecules that play a crucial role in gene regulation. In *Genetic Entropy,* Sanford calls DNA an instruction manual and writes, "There is no information system designed by man that can even begin to compare to the sophistication and complexity of the genome."[58]

Microsoft founder Bill Gates agrees with Sanford, saying, "DNA is like a computer program but far, far more advanced than any software ever created."[59]

Frankly, it is inconceivable to think that DNA is more complex and more elegant than the most sophisticated computer software written by human beings, by far, and it could have formed due to a series of unplanned and undirected processes. DNA forms the operating system for living organisms. Given the fact that human minds create computer programs, wouldn't it be logical to assume that a much more intelligent mind must be responsible for the existence of DNA?

Origin-of-Life Research Isn't Making Any Real Progress

Dr. James Tour is an organic chemist with a list of accomplishments so long it doesn't fit on a single page. He's an extremely competent, professional scientist and inventor who investigated the progress of origin-of-life research and challenged the claims of progress as being significantly exaggerated. Despite his Jewish heritage, Dr. Tour is an outspoken Christian, which makes him a target for atheist

58 Sanford, John. *Genetic Entropy and the Mystery of the Genome*. Third Edition. Page 7. New York. FMS Publications. 2008. Print.

59 Gates, Bill. *The Road Ahead*. Revised Edition. page 228. Viking. Penguin Group. New York. 1996.

detractors. During his debate with "Professor Dave" Farina, Dr. Tour was asked by an audience member to denounce the thinking of his followers to conclude that his criticisms of origin of life research served as evidence for God. Tour honestly replied, "I am telling you that just because we don't have an explanation for how life originated, I can't, as a scientist, say, therefore God made it magically happen and poofed it into existence."[60]

Atheists (such as Professor Dave) have often demanded proof for the existence of God while knowing very well that proof is an unattainable goal...or it has been until recently. With convenient timing for the purposes of being mentioned in this book, a Bulgarian scientist named Valentin Velchev has published an interesting paper titled "A Scientific Argument for the Existence of God"[61] that allegedly contains a mathematical proof for the existence of God.

Mathematics is the only discipline that offers to provide "proof" through its theorems. His argument is logical and compelling. Whether or not his math withstands scrutiny from the people whose work Velchev cited in his paper remains to be seen. While the evidence for God that Velchev mentions is largely circumstantial, it is interesting because the improbability of the Big Bang and the improbability of cosmic inflation aggregate with the improbability of abiogenesis, just as I have suggested. The unplanned universe gets increasingly improbable over time. Also, the alternative as a creative force to an intelligent Creator and a planned universe is unintelligent Time and random chance in an unplanned universe.

---

60 Tour, James. "Professor Dave Explains vs. Dr. James Tour", May 30, 2023, https://www.youtube.com/shorts/jNd5wrhGAyM

61 https://burevestnik-bg.com/книги/Argument-for-the-Existence-of-God.pdf

In any event, Velchev's paper will finally give skeptics and critics a comprehensive argument to critique.

In the "debate" between Dr. Tour and Professor Dave, Dr. Tour said that it was not his business to tell people what to think, and "Professor" Dave responded by accusing him of being a liar about science (even though Dr. Tour is a professional scientist and college professor while Dave is a YouTube personality with a BA in chemistry, an MA in science education, and a large internet following.) Dave's demeanor throughout the poorly moderated debate was antagonistic—he was obviously trying to provoke Dr. Tour into losing his temper to make him look bad. Dr. Tour has been a frequent critic of origin-of-life research and Dave apparently doesn't share his opinions on the subject.

After his debate with Professor Dave, Dr. Tour seems to be motivated to vindicate his position on origin-of-life research. He challenged ten top origin-of-life researchers to answer even one of five problems he had for them, promising he would remove all his online criticisms of their research he has published if they could answer even one of these five challenges:

1. Describe the origin of polypeptides.

2. Describe the origin of polynucleotides.

3. Describe the origin of polysaccharides.

4. Explain the origin of specified information contained in living things.

5. Create a living cell.

The origin-of-life researchers who received the challenge included Steve Benner, Jack Szostak, Clemens Richert, Bruce Lipshutz, Nicholas Hud, Lee Cronin, John Sutherland, Matthew Powner, Neal Devaraj, and Ramanarayanan Krishnamurthy.

Benner, Szostak, and Richert were selected to be the judges. Given Dr. Tour's vocal criticisms of origin-of-life research and the ability to forever silence him, answering at least one of those five questions for these ten researchers must have been a powerful motivation to respond to the challenge, yet none of them stepped up and accepted.

ANALYSIS OF THE ORIGIN OF LIFE

George Whitesides, winner of the 2007 Priestley medal, said about the cell in his acceptance speech,

> *"I believe that understanding the cell is ultimately a question of chemistry and that chemists are, in principle, best qualified to solve it. The cell is a bag—a bag containing smaller bags and helpfully organizing spaghetti—filled with a Jell-O of reacting chemicals and somehow able to replicate itself. Yes, it is important to know the individual reactions that make the cell what it is, but the bigger problem is understanding why life—the cell—is dynamically stable as a strongly interconnected network of reactions, organized in space and time in ways we do not grasp. Although we presently have no theory to explain this kind of system, understanding the kinetics of systems of coupled reactions is the kind of thing that chemists and chemical engineers are—in principle—uniquely qualified to do."*[62]

Is a living cell just a bag of chemicals that somehow manages to self-replicate, or is a cell a miniature factory with very complex machinery that performs important work which may be vital to the success of a much larger organism? Whitesides is probably

---

62 Whitesides, George. "Revolutions in Chemistry: Priestley medalist George Whitesides' address." Chemical and Engineering News, March 26, 2007. https://pubsapp.acs.org/cen/coverstory/85/8513cover1.html?

correct that chemists may be the most qualified to answer these types of questions, but that doesn't mean they can answer these questions. Dr. Whitesides also said,

> *"This [origin of life] problem is one of the big ones in science. It begins to place life, and us, in the universe. Most chemists believe, as do I, that life emerged spontaneously from mixtures of molecules in the prebiotic Earth. How? I have no idea. Perhaps it was by the spontaneous emergence of "simple" autocatalytic cycles and then by their combination. On the basis of all the chemistry that I know, it seems to me astonishingly improbable. The idea of an RNA world is a good hint, but it is so far removed in its complexity from dilute solutions of mixtures of simple molecules in a hot, reducing ocean under a high pressure of CO_2 that I don't know how to connect the two. We need a really good new idea. That idea would, of course, start us down the path toward systems that evolve autonomously—a revolution indeed."*[63]

What a profound expression of faith! Except Dr. Whitesides puts enormous faith in his own intellect and that of his fellow chemists instead of expressing faith in a superior intellect capable of creating the universe and life itself.

In his review of Carl Sagan's book *Demon Haunted World*, evolutionary biologist Richard Lewontin confessed,

> *"Our willingness to accept scientific claims that are against common sense is the key to an understanding of the real struggle between science and the supernatural. We take the side of science in spite of the*

---

63 Whitesides, George. "Revolutions in Chemistry: Priestley medalist George Whitesides' address." Chemical and Engineering News, March 26, 2007. https://pubsapp.acs.org/cen/coverstory/85/8513cover1.html?

> *patent absurdity of some of its constructs, in spite of its failure to fulfill many of its extravagant promises of health and life, in spite of the tolerance of the scientific community for unsubstantiated just-so stories, because we have a prior commitment, a commitment to materialism. It is not that the methods and institutions of science somehow compel us to accept a material explanation of the phenomenal world, but, on the contrary, that we are forced by our a priori adherence to material causes to create an apparatus of investigation and a set of concepts that produce material explanations, no matter how counter-intuitive, no matter how mystifying to the uninitiated. Moreover, that materialism is absolute, for we cannot allow a Divine Foot in the door."*[64]

Religious people are often accused of bias. But how much more biased can a person get? Lewontin is not the only famous scientist openly expressing his bias toward God by a long shot. George Wald basically said the same thing to *Scientific American* in his 1954 article "The Origin of Life." At a more recent gathering of chemists, biologists, and astronomers, geobiologist Ken Nealson admitted, "Nobody understands the origin of life. If they say they do, they are probably trying to fool you. In addition, they don't even know how to look for it."[65]

Abiogenesis is virtually impossible to understand on its own merit. In the context of a highly improbable universe produced by the Big Bang and cosmic inflation, the unplanned universe

---

[64] Lewontin, Richard C. "Billions and Billions of Demons", New York Review of Books. Page 28. January 9, 1997. https://www.nybooks.com/articles/1997/01/09/billions-and-billions-of-demons/

[65] Britt, Robert Roy. "The Search for the Scum of the universe," Free Republic. May 21, 2002. https://freerepublic.com/focus/chat/690927/posts

scenario presents too many difficulties to resolve. Molecular biologist Eugene Koonin writes,

> "The origin of life is one of the hardest problems in all of science, but it is also one of the most important. Origin of Life research has evolved into a lively, interdisciplinary field, but other scientists often view it with skepticism and even derision. This attitude is understandable and, in a sense, perhaps justified, given the 'dirty' rarely mentioned secret: Despite many interesting results to its credit, when judged by the straightforward criterion of reaching (or even approaching) the ultimate goal, the origin of life field is a failure—we still do not have even a plausible coherent model, let alone a validated scenario, for the emergence of life on Earth. Certainly, this is not due to lack of experimental and theoretical effort, but to the extraordinary intrinsic difficulty and complexity of the problem. A succession of exceedingly unlikely steps is essential for the origin of life, from the synthesis and accumulation of nucleotides to the origin of translation; through the multiplication of probabilities, these make the final outcome seem almost like a ***miracle.***"[66]

Like a miracle or an actual miracle, as in an act of God?

---

[66] Koonin, Eugene. *The Logic of Chance: The Nature and Origin of Biological Evolution*. Page 391. FT Press. Upper Saddle River, NJ. 2011.

Directed Panspermia

DNA co-discoverer Francis Crick calculated the time it would take for DNA to randomly form on Earth due to explicable natural processes and quickly determined the Earth wasn't old enough and there wasn't enough time to explain the existence of DNA. Crick's solution was to move the problem off Earth and assume DNA had some extraterrestrial origin. Specifically, Crick believed an alien form of intelligence might have created life and deliberately seeded Earth. It's difficult to gauge how serious Crick was about his hypothesis. It sounds far-fetched because it is far-fetched. No evidence supports it other than the belief that life appeared on Earth much sooner than it should have, assuming the origin was due to purely natural processes.

The problem with directed panspermia is that it merely removes the problem of the origin of life away from Earth. The theory proposes an intelligent origin of life, but the source of intelligence is still subject to the problems of the successful Big Bang and cosmic inflation because the source of intelligence is thought to be an alien civilization that exists within the confines of the universe. According to John Lennox, Fred Hoyle agreed with Francis Crick.

Lennox said: "One of my examiners at Cambridge was Sir Fred Hoyle, and he came to Cardiff, where I was a lecturer many years ago, and he shocked everybody. He just stood up and said to an absolutely packed crowd, because he was famous, he said, "Life cannot have originated on earth," and there was a collective gasp. And he said,

> *"I've done the calculations, and mathematically it is simply impossible. There isn't enough time. And I actually have a copy of those calculations at home. And he just said it's quite obvious if you do the calculations, and he puts it very simply and pregnantly. Rabbits produce rabbits and very little else. And what his mathematics, I think, showed him was that the*

> *innocent aspect of evolution, which we can all accept, that is, you get minor variations on a theme, which Michael here has dealt with so successfully in his book The Edge of Evolution. That's non-controversial. But once you go beyond that and think of new animals, new body plans, all of that kind of thing."*[67]

Planned Versus Unplanned

Mathematical calculations have been performed by scientists such as Francis Crick and Fred Hoyle to estimate the odds against the success of a living organism being created out of inanimate matter, and the results have not been pretty. Francis Crick tried to calculate how long DNA would take to self-assemble and then hypothesized that life formed in outer space and perhaps had been sent to Earth by extraterrestrials to seed life on this planet because his calculations showed that life appeared much sooner than his estimates suggested was possible.

Darwin originally speculated that life might have begun in a warm, quiet pond, but those ponds simply did not exist on prehistoric Earth. Shortly after the Earth formed, its surface was an inhospitable cauldron, not a warm, quiet pond. In a letter, Charles Darwin wrote,

> *"It is often said that all the conditions for the first production of a living organism are now present, which could ever have been present. But if (& oh what a big if) we could conceive in some warm little pond with all sorts of ammonia and phosphoric salts—light, heat, electricity etc. present, that a protein compound was chemically formed, ready to undergo still more*

---

[67] Lennox, John. "By Design: Behe, Lennox, And Meyer On The Evidence For A Creator -- Uncommon Knowledge with Peter Robinson", Hoover Institution, October 15, 2022, https://www.youtube.com/watch?v=rXexaVsvhCM&t=3s

> *complex changes, at the present day such matter would be instantly devoured, or absorbed, which would not have been the case before living creatures were formed."*[68]

Wishful thinking.

Abiogenesis Versus Spontaneous Generation

What you might think about abiogenesis and spontaneous generation will almost certainly depend on whom you ask. According to the website, Biology Stack Exchange, abiogenesis happens very infrequently, but spontaneous generation occurs all the time, as new species form from existing species. However, another website, Science Professor Online, claims that spontaneous generation is a debunked idea, suggesting life can, on a virtually daily basis, arise from inanimate matter. So, which argument is correct?

Louis Pasteur performed an experiment which proved the spontaneous generation hypothesis was false. First, he observed that flies seemed to appear spontaneously on rotting meat and devised an experiment where he boiled a meat broth in a flask and let it sit with a filter covering it. He also boiled a broth but left it uncovered and discovered bacteria growing in the broth. The experiment showed that organisms growing in such broths came from external sources rather than being spontaneously generated.

Abiogenesis, the unique creation of a living organism from inanimate matter, logically must have occurred, otherwise life would not exist. The problem is the origin-of-life research hasn't made any genuine progress toward replicating the creation of a living cell from inanimate matter.

---

68 Darwin, Charles. "Letter to J.D. Hooker." February 1, 1871.

RNA World Hypothesis

The RNA World hypothesis is the idea that RNA molecules evolved first, and DNA evolved from RNA. The problem with this idea is that DNA has a different composition than RNA, with uracil in RNA replacing thymine in DNA. However, if uracil undergoes a chemical reaction, it turns into cytosine, not thymine. In other words, you can't go from RNA to DNA by chemical reaction; the process requires a chemical reconstruction.

Science educator Arvin Ash said,

> *"The origin of living organisms from inorganic or nonliving material is called abiogenesis. It's important to distinguish this from evolution. Abiogenesis is not evolution. Evolution is the process of development or the diversification of living things from earlier forms of living things. Evolution does not say anything about how life first originated. So, how did the first life originate? Despite the incredible variations that we see today at the fundamental level, all living things contain a trinity of elements. First, nucleic acids, which make up the DNA or its simpler form called RNA. These contain the blueprints of life and are self-replicating molecules. Second, there are proteins, which are the workhorses that perform the important functions of your body. And third, there are lipids, which encapsulate the cells of your body. Before any living things existed, before animals, plants, and even bacteria existed, these three things had to have been present in the primordial soup in order for life to start. Some argue that the most important components of this trinity are the lipids that make up the cell walls. Why would this be the most important? Because without the wall, or a way to encapsulate certain elements within the soup, there would just be a soup of material that would just be disorderly and floating around in a sea*

of liquid. It would not be functioning inside something that could potentially self-replicate."[69]

Fair enough. Defenders of evolution theory like to point out that evolution is a question for biology and abiogenesis is a question for chemistry. In other words, non-overlapping magisteria. However, as I have often pointed out, life cannot evolve until it exists. From LUCA, created by abiogenesis, there must be a continuum from the earliest to modern life, if the unplanned universe theory is true. There can be no interruptions.

Ash continues,

> *"Where do lipids come from? It was once thought that they could only be produced by living cells. But experiments have shown that when carbon monoxide and hydrogen are heated up with minerals commonly found in Earth's crust, lipids can form. All components were available on the early Earth and could have happened in underwater hydrothermal vents. You might at this point say, 'Aha! That's it!' That's how the first cell must have formed. Not so fast. It turns out that while lipids do have this quality of self-assembly when there are certain ions present, such as salts or magnesium, it destroys the lipid structure. They disintegrate. But the problem is that RNA and other functions of a cell require these ions, and since the early Earth was believed to have salty oceans, and*

---

[69] Ash, Arvin. "How did life begin? Abiogenesis. Origin of life from nonliving matter", Arvin Ash, September 6, 2019, https://www.youtube.com/watch?v=nNK3u8uVG70&t=352s

since these spheres can't form in these salty oceans, this theory always had a gaping hole."[70]

The idea that RNA formed first, and DNA evolved from RNA, is called the RNA World Hypothesis.[71] Because DNA is a more complex molecule that contains three of the same four components as RNA, it has become somewhat popular to believe that RNA formed first, and DNA formed afterward. But as we just saw moments ago, RNA can't become DNA with a chemical reaction. Because we know the universe has not always existed, we can safely assume that life has not always existed. In the unplanned universe hypothesis, life formed by some unknown process for some unknown (or nonexistent) reason. Darwin envisioned life beginning in a warm, gentle pond. More contemporary speculation among biologists favors the first life form being some kind of extremophile, from the taxonomic domain Archaea. No matter which is true, in the unplanned universe, luck was not only involved, it was instrumental.

There are two basic kinds of cells: prokaryotes and eukaryotes[72]. All cells have cell membranes, cytoplasm, ribosomes, and DNA. Prokaryotes are all single-celled organisms that reproduce asexually. The cell simply copies its DNA and then splits in half to form two cells, in a process known as binary fission. Thus, offspring are identical to their parents. Eukaryotes are far more complex cells that contain many more organelles (a cellular

---

70 Ash, Arvin. "How did life begin? Abiogenesis. Origin of life from nonliving matter", Arvin Ash, September 6, 2019, https://www.youtube.com/watch?v=nNK3u8uVG7o&t=352s

71 Ash, Arvin. "How did life begin? Abiogenesis. Origin of life from nonliving matter", Arvin Ash, September 6, 2019, https://www.youtube.com/watch?v=nNK3u8uVG7o&t=352s

72 Scoville, Heather. "Learn About the Different Types of Cells: Prokaryotic and Eukaryotic", ThoughtCo, Apr. 5, 2023, https://www.thoughtco.com/types-of-cells-1224602.

subunit that performs a specific function). Some examples of these organelles include lysosomes, mitochondria, and the Golgi apparatus. Most organisms with eukaryotic cells form sexual partnerships between males and females to produce offspring that are not identical to their parents. Prokaryotes cannot be seen with the naked eye and must be viewed under a microscope. Every visible plant and animal on Earth is composed of billions of eukaryotic cells that perform specific functions to serve the whole organism. Heart cells form an organ called the heart that keeps blood pumping throughout the organism, while liver and kidney cells form other unique organs that similarly perform specific tasks serving the whole organism.

However, every cell in the organism has identical DNA, so the question becomes, how does an individual cell know what type it should become? How do kidney cells know to become kidney cells, and brain cells to become brain cells? The process is known as differentiation.

Science writer John Staughton wrote, "Most experts agree that unicellular life arose 4.1-3.5 billion years ago, while the first complex form of multicellular life first formed around 600 million years ago. Generally, it is believed that unicellular life reigned supreme for more than 2 billion years before the evolution and spread of multicellularity."[73]

But that doesn't explain how individual cells know what type of cells they need to become, does it?

The progression imagined for the so-called Tree of Life that began with LUCA (last universal common ancestor) and progressed until today is the view of the unplanned universe.

---

[73] Stoughton, John. "How Long Did It Take for Multicellular Life to Evolve from Unicellular Life?", Science ABC, October 19, 2023, https://www.scienceabc.com/pure-sciences/how-long-did-it-take-for-multicellular-life-to-evolve-from-unicellular-life.html.

Dr. Rob Stadler has a PhD in medical engineering from Harvard University and has patented more than 140 medical devices in his career spanning two decades. Rather than a Tree of Life, Dr. Stadler proposes a whole Forest of Life where multiple organisms share a common designer as opposed to common descent. Internet personalities like "Professor Dave", a woman calling herself "Gutsick Gibbons," and a PhD who identifies as "Creation Myths" gathered online to disparage Dr. Stadler for daring to question the orthodoxy of the Tree of Life and universal common descent.

Interestingly, advocates of abiogenesis will be quick to say that whatever process turned non-life into living organisms must have produced more than one living cell. A small population of cells is necessary for life to survive and persist. But the Tree of Life (and more importantly, abiogenesis experiments) imply only one LUCA is possible.

Chris Impey writes in his book *The Living Cosmos*,

> *"Chemists tell a joke among themselves, when nobody else is around. Life is impossible, they say: we've put simple chemical ingredients in water, we've added energy in the form of heat or electricity, and all we ever get is an organic sludge. We never see replicating molecules. We never make a cell. The astronomer Fred Hoyle once said that the act of assembling the simplest living organism from simple molecular ingredients was as unlikely as a tornado whipping through a junkyard and assembling a jumbo jet. Yet somehow it happened. Was it blind luck? And if it happened here, could it happen somewhere else?"*[74]

---

[74] Impey, Chris. *The Living Cosmos*. Page 73. New York. Random House. 2007. Print.

Impey openly admits to being an atheist. His honesty is refreshing. He admits that we don't even have a scintilla of evidence showing life could self-organize into a living cell, yet we have life, so abiogenesis must be possible. Once again, the wisdom of Fred Hoyle is quoted. Also, Impey frames the question correctly—if not made by God, is luck responsible for the origin of life?

In his book *The Way of the Cell,* molecular biologist Franklin Harold wrote,

> *"Of all the unsolved mysteries remaining in science, the most consequential may be the origin of life. This opinion is bound to strike many readers as overblown, to put it mildly. Should we not rank the Big Bang, life in the cosmos, and the nature of consciousness on at least an equal plane? My reason for placing the origin of life at the top of the agenda is that resolution of this question is required in order to anchor living organisms securely in the real world of matter and energy and thus relieve the lingering anxiety as to whether we have read nature's book correctly. Creation myths lie at the heart of all human cultures, and science is no exception; until we know where we come from, we do not know who we are. The origin of life is a stubborn problem, with no solution in sight. There is indeed a large and growing literature of books and articles devoted to this subject, many with theories to propound. Biology textbooks often include a chapter on how life may have arisen from non-life, and while responsible authors do not fail to underscore the difficulties and uncertainties, readers still come away with the impression that the answer is almost within our grasp. My own reading is considerably more reserved. I suspect that the upbeat tone owes less to the advance of science than to the resurgence of primitive religiosity all around the globe, and particularly in the West. Scientists feel vulnerable to the onslaught of*

> *believer's certitudes, and so we proclaim our own. In reality, we may not be much closer to understanding genesis than A.I. Oparin and J.B.S. Haldane were in the 1930s, and in the long run, science would be better if we said so. After all, the unique claim of science is not that it has all the answers, but that it knows the questions and will not compromise its commitment to the rational search for truth."75*

Dr. Harold's assessment of the progress of origin-of-life research seems to align with Dr. Tour's—despite the impression that scientists are on the verge of creating life in the lab, the reality is that no one is even close. However, Dr. Harold should not be mistaken for a theist. He seeks a purely natural explanation for the inexplicable. A few pages later, Harold adds,

> *"We (cellular biologists) are compelled by our calling to insist at all times on strictly naturalistic explanations: life must, therefore, have emerged from chemistry. Granted also that simple organic molecules were present at the beginning, in uncertain locations, diversity and abundance. Leave room for contingency, some rare chemical fluctuation that may have played a seminal role in the inception of living systems and remember you may be mistaken. With all that, I still cannot bring myself to believe that rudimentary organisms of any kind came about by the association of prefabricated organic molecules, born of purely chemical processes in their environment."*[76]

For life to arise by purely naturalistic chemical processes, a lot of luck would need to be involved to explain how the perfect

---

75 Harold, Franklin. *The Way of the Cell: Molecules, Organisms, and the Order of Life*. Page 235. New York. Oxford University Press. 2001. Print.

76 Harold, Franklin. *The Way of the Cell: Molecules, Organisms, and the Order of Life*. Page 250. New York. Oxford University Press. 2001. Print.

combination of chemicals managed to coalesce and react until a living organism magically poofed into existence, as Dr. Tour suggested. The wrong amino acids or amino acids with the wrong chirality would cause the origin of life to fail.

Harold observes, "It would be agreeable to conclude this book with a cheery fanfare about science closing in, slowly but surely, on the ultimate mystery; but the time for rosy rhetoric is not yet at hand. The origin of life appears to me as incomprehensible as ever, a matter for wonder but not for explication."[77]

DNA

DNA is encoded information. As Bill Gates said, DNA is considerably more complex than any computer program ever devised by humans. DNA is a recipe for a living organism, with the baking instructions also coded into the message. A computer's operating system is composed of hundreds of millions of bytes of information stored on disk and occupies the lowest level memory of the computer when the computer boots up. Comparatively, billions of bits of information are stored within the tiny confines of a living cell.

Frank Turek explained the significance of DNA, saying,

> *"If you're walking along the beach and see in the sand 'John loves Mary', you don't assume the waves did that, or crabs came out of the water and made that message. You know that message, even a short message like 'John loves Mary' requires a mind. Natural forces don't create messages. And so, a short message like that, if that requires a mind, how about a message three point two billion letters long, in every one of our forty trillion cells? That's got to come from*

---

[77] Harold, Franklin. *The Way of the Cell: Molecules, Organisms, and the Order of Life*. Page 251. New York. Oxford University Press. 2001. Print.

a mind as well. Even Bill Gates, the founder of Microsoft, said that DNA is like a computer program, but it's far, far more advanced than any software we've ever created. Well, where I come from, if there's a program, there has to be a programmer. If there's a code, there's got to be a coder. If there's a message, there's got to be a messenger. What we see in DNA is like binary code. Binary code only comes from a mind. It doesn't come from an impersonal force. It doesn't come from the four natural forces of nature: strong and weak nuclear forces, electromagnetism, and gravity. You don't get messages from those things. And every living thing has a message. Even the simplest living thing—an amoeba has at least one thousand volumes worth of an encyclopedia of information in it. Even Richard Dawkins admits that. But where does that come from? It can't come from natural forces. It comes from a mind...We're not arguing from a gap in our knowledge. When you see 'John loves Mary' on the sand, you don't just lack a natural explanation for that. You don't just have a gap in your knowledge about natural forces. It's that 'John loves Mary' is positive, empirically verifiable evidence for an intelligent cause. The same thing is true for our genetic code. So, we're arguing from what we do know, not what we don't know."[78]

Okay. Turek has a strong opinion, but he's not really an expert on DNA (nor am I, for that matter). John Lennox also had this to say:

"The human genome is the longest word we've ever discovered. And we can call it a word because it's

---

[78] Turek, Frank. "Does DNA Point to a Designer? Watch This!", Cross Examined, April 21, 2021, https://www.facebook.com/CrossExamined.org/videos/does-dna-point-to-a-designer-watch-this/880442446070138/

> *written in a chemical language of four letters. And all those letters, strung out like a computer program, have got to be in the right order, otherwise it breaks down. Most programs, if you change a letter, that's the end of the program. So, we're dealing with something absolutely gigantic in terms of probabilities before we even think of the extra complexity that arises through the folding of the proteins and all the epigenetic information that's been discovered in recent years."*[79]

Lennox is a mathematician, not a geneticist, but that doesn't mean he knows nothing about DNA. We've only known that DNA exists for less than 100 years. We've barely scratched the surface in terms of what there is to know about it. Assumptions about DNA that have been made within my lifetime have been proved incorrect. There are no "junk" sequences in DNA. The purpose of a sequence of DNA may currently be unknown, but it should never be assumed to be an "ancient repetitive element." The common denominator of DNA is that it is always composed of the same four nucleotides that form a pattern unique to the organism in question: adenine, thymine, cytosine, and guanine. Every living organism on Earth, both plant and animal alike, has DNA composed of specific sequences of those same four nucleotides.

In that same program featuring John Lennox and Michael Behe, Stephen Meyer added,

> *"The 97% that doesn't code for proteins they said is just the leftover the flotsam and jetsam of the random trial and error process of natural selection and random mutation. Several of our leading ID proponents in the 1990s and early 2000s said, "Well,*

---

79 Lennox, John. "By Design: Behe, Lennox, And Meyer On The Evidence For A Creator -- Uncommon Knowledge with Peter Robinson", Hoover Institution, October 15, 2022, https://www.youtube.com/watch?v=rXexaVsvhCM&t=3s

> *we agree that there should be some evidence of those random mutations accumulating, but not 97%, so we're going to predict, based on our theory, that those non-coding regions of the DNA aren't junk DNA, but rather they are importantly functional." And that prediction has now been confirmed by the Encode Project and a number of other interesting developments in bioinformatics. And we now know that the non-coding regions of the DNA are functioning roughly like an operating system in a computer. They're controlling the timing and expression of the coding files. This is a prediction that was made by proponents of intelligent design that's been confirmed by new discoveries."*[80]

We have evidence that DNA that doesn't code for protein nevertheless has a functional purpose. Therefore, it isn't junk DNA. It's part of the unique code that defines an organism.

In another interview, Meyer said,

> *"The big discovery of the 1950s and 1960s was that the DNA molecule encodes information in a roughly digital or alphabetic or typographic form. In computer science we have characters you know, zeros and ones. This is (Francis) Crick in 1957. It's the sequence hypothesis that he realized the information in DNA or the chemical subunits of DNA called nucleotide bases were functioning like alphabetic characters in a written text or like the zeros and ones in a section of computer code. That is to say it's not—it wasn't their chemical properties that gave them their function but their specific arrangement in accord with an*

[80] Meyer, Stephen. "By Design: Behe, Lennox, And Meyer On The Evidence For A Creator -- Uncommon Knowledge with Peter Robinson", Hoover Institution, October 15, 2022, https://www.youtube.com/watch?v=rXexaVsvhCM&t=3s

> *independent symbol convention which was later explicated in the form of what we call the genetic code. So, we had genetic text functioning according to a code."*[81]

DNA is like a code because it *is* a code. Code processes data to produce information, much like a computer does. The code for human DNA produces a human being. The genetic information will vary, and the product will vary depending on the input for sex, eye color, hair color, height, and other characteristics, but the result will always be a human being. Meyer also said,

> *"This is a genuine information storage system. Crick, by the way, was a code breaker in World War II. So, this is fascinating, is application of information sciences to molecular biology. And this is the argument that I make is that what we know from experience is that information, whether we find it in a hieroglyphic inscription or a paragraph in a book or information embedded in a radio signal or in a section of computer code—whenever we find information and we trace it back to its ultimate source we always come to a mind, not a material process. And what I do in the book* Darwin's Doubt *and my prior book* Signature in the Cell *was show that these* ***undirected*** *evolutionary mechanisms that have been proposed as an explanation for the origin of information fail for various reasons. We've talked about the reason the Darwinian mechanism fails is because it can't search the space when it's so vast that the odds are overwhelmingly against it. So, if we, from a materialistic evolutionary standpoint, don't have any*

---

[81] Meyer, Stephen. "Darwin DEBUNKED: Using Breakthroughs in Math and Science", Daily Dose of Wisdom, April 24, 2024,

https://www.youtube.com/watch?v=e8U8QV8HNDg

> *explanation for the origin of the information that's necessary to build a new biological form. And yet we do know from our uniform and repeated experience, which is the basis of all scientific reasoning, of a source of information, of a cause of the origin of information—that cause is intelligence or mind. What I've argued in both* Darwin's Doubt *and* Signature in the Cell *is that what we're seeing in life is evidence the activity of a directing mind in the history of life."*[82]

Meyer made the point that evolutionary mechanisms and processes are theoretically undirected, which I've also said, but with less-fancier language. If you take a living organism and try to create a new organism via sexual reproduction, we'll need a reasonable explanation for the source of the new genetic information.

CONCLUSION

The Bible doesn't have much to say about the creation of individual cells. It primarily speaks of God creating living organisms in a variety of forms. It does not describe more complex forms evolving from simpler forms over long periods of time. Interestingly, the Bible speaks of God making creatures invisible to the naked eye thousands of years before the invention of the microscope by Hans and Zacharias Jansen around the year 1590.

Colossians 1:16 says,

> *"For by Him all things were created, both in the heavens and on earth,* ***visible and invisible****, whether*

[82] Meyer, Stephen. "Darwin DEBUNKED: Using Breakthroughs in Math and Science", Daily Dose of Wisdom, April 24, 2024,

https://www.youtube.com/watch?v=e8U8QV8HNDg

> *thrones or dominions or rulers or authorities—all things have been created through Him and for Him."*

Even though the Bible doesn't mention bacteria by name, it does speak of invisible life.

In an unplanned universe, plants and animals simply exist because they can exist. Earthworms do not exist to feed birds or perform the essential services they provide for maintaining soil health. The only reason worms exist is because unplanned, undirected, and unknown forces caused the worm to exist for no reason. Conversely, in a planned universe, earthworms were created by God to provide benefits to the Earth and to animals that like to eat worms, such as birds. In a planned universe, things like food chains make sense because they must have been created for some reason. In an unplanned universe, food chains don't make quite as much sense because there is no explanation for the emergence of order from chaos. The Bible makes it clear that God created different kinds of living organisms to serve a specific purpose, such as providing food for other organisms. In other words, a food chain.

Psalm 104:14-15 reads,

> *"He makes grass grow for the cattle, and plants for people to cultivate—bringing forth food from the earth: wine that gladdens human hearts, oil to make their faces shine, and bread that sustains their hearts."*

Grass is food for cattle. Cows and other plants become food for people. Cows give us milk, which can be drunk or made into cheese by curdling the milk.

In Romans 4:17, Paul writes,

> *"As it is written: I have made you the father of many nations—in the presence of the God in whom he*

> *(Abraham) believed,* ***the one who gives life to the dead and calls things into existence that did not exist.*** *"*

So, the Bible is claiming that God brought matter into existence and created life out of inanimate matter. Not too shabby for a group of illiterate shepherds who allegedly knew nothing about science, right? Just kidding. If they were really illiterate, they couldn't have written the Bible, for one thing. But why did people believe the universe had a beginning thousands of years before the evidence for the Big Bang was discovered?

In the book of Genesis, the Bible clearly describes a Forest, not only one Tree of Life. However, in the book of Revelation, Chapter 22 verse 18-19 reads,

> *"I warn everyone who hears the words of prophecy of this book: if anyone adds to them, God will add to him the plagues described in this book, and if anyone takes away from the words of the book of this prophecy, God will take away his share in the* ***tree of life*** *and in the holy city, which are described in this book."*

So, which is it? Is life a forest or a tree? Of course, forests are composed of trees. In a pine forest, you'll typically find rows and rows of pine trees planted in an organized pattern to make cutting down and removing the trees easier, but no other kinds of trees will be found there because humans planted the pine trees as a crop and plan to harvest the wood. They manage the forest. But in a natural forest, you'll find many kinds of trees with haphazard spacing.

According to the unplanned universe theory, the Big Bang, cosmic inflation, and now the origin of life are merely fortuitous ("lucky") sequences of events that culminate with the creation of you. In the theistic evolution variation, God got lazy after creating the initial life form and allowed evolution to run its course from that point forward. In the theistic worldview, God created the universe, and God created life. That is the simplest, most straightforward answer.

It is my understanding that people who believe in an unplanned origin of life believe that a single individual organism could not have survived—it has been assumed that life must have spawned within a population rather than a single individual cell. By the same token, we have been told that abiogenesis is extraordinarily difficult and extremely rare. Success, even once, has not been credibly explained. So, why should anyone believe multiple organisms spontaneously transformed from the inanimate to animated form in roughly the same timeframe? It makes sense to assume a small population rather than an individual organism was necessary for the sake of survival, but abiogenesis should be even more difficult for a group than for an individual.

[illegible] universe, and therefore not life? That is the simplest [illegible] most straightforward answer.

[illegible]

MIRACLE 4
THE DIVERSITY OF LIFE

INTRODUCTION

Let's ask the million-dollar question right out of the gate: Does the evidence for the theory of evolution somehow disprove or eliminate the need for creation? No, it most certainly does not. How can I be so confident? Because the "scientific theory" of evolution and the philosophy of creationism do not directly compete against each other. Life cannot evolve until it exists. Period. Exclamation mark, even. Long before evolution ever becomes possible, at least two creation events and a third miracle have already occurred: the origin and expansion of the universe and the origin of life. Even so, there is no reason to believe that wildly different creatures, such as humans and octopuses, share a common ancestor through sexual reproduction over long periods of time. However, if universal common descent is true, we are related after lots of sexual reproduction over very long periods of time. Nevertheless, there is no logical or rational reason, nor any genuine evidence, to support the belief that humans might be related to fish or octopuses through descent via sexual reproduction over long periods of time. Furthermore, there isn't as much time as we might initially guess, given the age of the Earth, because of evidence for multiple mass extinctions found in the fossil record.

On the other hand, if we believe in the creative power of God, does that mean God created bluebirds and blue jays, cardinals and crows, sparrows and vultures individually? Or could God have simply created the blueprint for a kind of animal, in this instance a bird, and then allowed His creation to "go forth and multiply" into a wide variety that "evolved" into multiple species? I don't have a problem with that idea. And the word evolution simply means "change." But it is not undirected and unplanned change, so it should be called intelligent design, not evolution. Universal common descent requires shape-shifting to be true.

In fact, it is more plausible to believe God created the bird and then allowed 18,000 varieties of birds to self-organize by natural selection. Blue jays are naturally attracted to other blue jays, not cardinals. Of course, God didn't need to create each unique variety of bird, just the prototype. God created "fish" but perhaps not flounder, trout, salmon, mackerel, or cichlids individually. After all, the genome of a crow is vastly different than the genome of a turtle or the genome of a human being. Each genome will naturally have some nucleotide sequences in common, but enough of the genome will be sufficiently unique to determine one unique organism's DNA from another. If evolution is true, as Jerry Coyne claimed in the title of his most famous book, then we should believe every living organism on Earth is related to every other living organism because of sexual reproduction. Think about it...every plant, every animal, every insect, fish—if it's alive and universal common descent is true, it is literally related to every other living organism on Earth *because of sex*, with sufficient time. This idea of a universal relationship isn't terribly difficult to imagine when using comparative anatomy, as we look at similar organisms such as horses and zebras or humans and apes. We can observe similarities. However, when asked to apply the exact same premise to more diverse organisms such as cows and whales or humans and seahorses, our imagination must be stretched and contorted into far less comfortable positions.

Even so, we have been assured by the experts (Richard Dawkins, Jerry Coyne) that the theory has been proved beyond all doubt and there is no reasonable alternative. There is always an alternative, but perhaps not an alternative these experts like.

The obvious problem with the "settled science" logic is that this so-called unassailable theory absolutely depends on the success of an untested, unproven hypothesis called abiogenesis. Atheists prefer to argue that the origin of life is completely unrelated to the diversity of life, saying that we can observe evolution and decree it to be fact because abiogenesis and evolution aren't even in the same scientific fields—the former is in chemistry and the latter is in biology.

Life is a miracle. The idea that even one single-celled organism on Earth is alive is a miracle. In fact, that was the third of our seven miracles. The existence of all the living organisms on Earth is yet another miracle, now our fourth miracle. The two extreme alternatives we have been told we must choose between are the biblical account of special creation from the book of Genesis, where the Earth is allegedly only thousands of years old, OR Darwinian natural selection with billions of years, supported by science — all the so-called real evidence allegedly supports evolution. Paleontological and geological evidence allegedly supports an Earth roughly four billion years old, providing the necessary time for naturalistic evolution to occur.

Or does it? The fossil record also shows evidence of multiple mass extinctions, which compress the time available for evolution to occur. From studying the fossil record, a pattern called punctuated equilibrium appears. This pattern, identified by paleontologists Stephen Jay Gould and Niles Eldridge, suggests that rather than evolving slowly over long periods of time, most animals seem to suddenly appear in the fossil record, then remain virtually unchanged for a long period of geologic time called stasis, before going extinct due to a mass extinction event. Even though punctuated equilibrium seems to contradict the central premise of Darwinism, the people who conceived of the theory to explain the evidence have insisted there is no conflict between punctuated equilibrium and Darwinian evolution. It's almost as if a religion named Evolution popped into existence. Direct challenges to the orthodoxy are tantamount to heresy.

As described by everyone from Charles Darwin and Richard Dawkins to biologists Jerry Coyne, Neil Shubin, and Ken Miller, the process known as evolution is an *undirected* process. There is no room for the creativity of God. Time and Luck can explain any perceived improvements from "ancestor" to "descendant" species. Animals are not the result of an intelligent design, but the result of the blind forces that often produce the illusion of utility and design. According to evolutionary theory, birds do not have wings or hollow bones for the purpose of flying. They have wings and hollow bones as the result of a series of tiny mutations aggregated over an unknown (but long) number of years, without the goal of flight. Theistic evolution is a nonsense explanation because the Creator is excused from the very act of creation, to be replaced by Time and Luck as the unplanned and undirected brute forces responsible for all life on Earth.

If you want to believe in the theory of evolution, be my guest. I really don't care. I only ask that you don't claim evolution is the *only* plausible explanation for the diversity of life, because it isn't, or that evidence exists for evolution that doesn't exist for intelligent design—it is the exact same evidence with two different interpretations. Creation and intelligent design are the planned alternatives to unplanned evolution. The theory of evolution is a philosophy, not a scientific theory. The scientific method cannot be applied because almost anything can become possible with enough time and enough imagination. The problem with evolutionary theory is that it always wants to begin with a breeding population of creatures. There is never a "first" couple of new organisms; the scenario is always that a group of organisms splits apart and continues to reproduce as two separate populations. Over time, mutations and adaptations accumulate until, at some point, there are enough differences within the original group and the isolated second group that they can be called two distinct species. That description may not fully satisfy any biologists reading this book, but I challenge any reader to compare what Jerry Coyne has written about speciation to my generalization and point to any specific words that are untrue. For that matter, point to the untrue words in my next claim: if the secular theory for the diversity of life is to be true, the secular theory for the origin of life must also be true—that the first living organism formed (without assistance) from inanimate matter. That first organism multiplied via asexual reproduction, basically cloning itself.

Richard Dawkins wrote,

> *"Evolution is a fact. Beyond reasonable doubt, beyond serious doubt, beyond sane, informed, intelligent doubt, beyond doubt, evolution is a fact. The evidence for evolution is at least as strong as the evidence for the Holocaust, even allowing for eyewitnesses to the Holocaust. It is the plain truth that we are cousins of chimpanzee, somewhat more distant cousins of monkey, more distant cousins still of aardvarks and manatees, yet more distant cousins of bananas and turnips...continue the list as long as desired."*[83]

However, Dr. Marco Fasoli strongly disagrees with Dawkins:

> *"Evolution is not observed and is not happening today. It has not been observed in the past. It presupposes but cannot account for the origin of life. I'm talking primarily about biological evolution here. It violates an important physical law, the 2nd law of thermodynamics. Its proposed mechanism of natural*

[83] Dawkins, Richard. *The Greatest Show on Earth: The Evidence for Evolution.* Page 8. New York. Free Press. 2009. Print.

> *selection, and mutation has been shown conclusively not to work. It fails to account for the irreducible complexity that we observe in biological systems. It fails to account for the origin of specified information in the biological systems, primarily the genetic code. And it fails because it violates many first principles in philosophy which actually renders the very concept of evolution metaphysically impossible and hence irrational."*[84]

Some might argue that Dawkins is a biologist (zoologist) and Fasoli has his PhD in chemistry, which makes Dawkins more qualified to speak about evolution—but Fasoli's expertise in chemistry means he's the authority on the origin of life, so it's sort of a tradeoff. Life is the common denominator. Because the diversification of life depends completely on there being an origin of life, Fasoli's argument becomes a bit more plausible because intelligence is assumed to have an intelligent origin.

While appearing on a program called "Uncommon Knowledge" with Michael Behe and John Lennox, intelligent design advocate Stephen Meyer said,

> *"The Cambrian Explosion refers to an event in the history of life in which the major groups of animal forms, the new body plans that are exemplified by the largest categories of different types of animals, appear very abruptly in the fossil record with no discernible connection to ancestral precursors or intermediates in the lower pre-Cambrian strata. And this pattern of abrupt appearance of the major groups of organisms of biological or morphological innovation, as it's*

---

[84] Fasoli, Marco. "7 Scientific Reasons why Darwinian Evolution is a Myth", Radio Immaculata, April 23, 2024

https://www.youtube.com/watch?v=D9zL-f8lSZk

> *called, recurs up and down the sedimentary rock column, the first winged insects, the first dinosaurs, the first birds, the first mammals, the first flowering plants. There are multiple instances of this type of abrupt appearance, and so the fossil record looks very different than Darwin anticipated that it would look. He depicted the history of life as a great branching tree where the forms of life we see today emerged gradually from one or very few simple forms at the base of the tree, at the trunk of the tree. But instead, what we see, it looks more like a lawn or perhaps an orchard of separate trees where the major groups of organisms appear abruptly without connections to those ancestral precursor forms."*[85]

Later in the interview, Meyer also said,

> *"Well, he [Michael Behe] kind of has, he kind of has [made the flagellar motor famous]. But it's a 30-part rotary engine. The ATP synthase is a turbine with multiple parts. But the parts are made of proteins. And proteins are, in essence, the toolbox of the cell. They perform specific functions in view of their three-dimensional shapes. So, they make the parts of molecular machines, they function as enzymes to catalyze reactions at super-fast rates, they help process information. But if you were to build a system like the flagellar motor, you need 30 proteins that fit together in an integrated fashion. But that requires genetic code. Each one of those proteins requires a long stretch of genetic information to build the protein. And so, what you're talking about is not just some bent*

---

[85] Meyer, Stephen. "By Design: Behe, Lennox, And Meyer On The Evidence For A Creator -- Uncommon Knowledge with Peter Robinson", Hoover Institution, October 15, 2022, https://www.youtube.com/watch?v=rXexaVsvhCM&t=3s

> *hammer or something sitting around doing nothing, you're talking about a need for genetic information that is sufficient to overcome these long odds against building the protein in the first place. So, it'd be like, to change the metaphor slightly, a gigantic haystack the size of the North American continent, and you're only allowed to search one 37th of the continent, maybe a tiny little square of southern California. If that's the case, are you more likely or less likely to find the needle? And the answer is you're overwhelmingly more likely not to find the needle than to find it, which is to say the mutation selection mechanism lacks the creative power to generate new biological information."*[86]

The problem with the argument that life could build complexity gradually over long periods of time is the introduction of purpose while simultaneously denying its existence. The belief is that evolution started with an organism without eyes or without wings and developed a half of a wing or an eye that is barely functional as an interim step on the way to fully functioning wings and eyes that can see rather well. When scientists have performed experiments in the lab, it has been a practical experience for an organism to lose function, rather than gain it.

Still later in that same interview, Stephen Meyer added,

> *"Could this mystery of the origin of information actually be a positive indicator of a different type of cause at work altogether? And the person who helped me most think this through is actually Darwin himself. Because Darwin pioneered a method of historical*

---

[86] Meyer, Stephen. "By Design: Behe, Lennox, And Meyer On The Evidence For A Creator -- Uncommon Knowledge with Peter Robinson", Hoover Institution, October 15, 2022, https://www.youtube.com/watch?v=rXexaVsvhCM&t=3s

scientific reasoning where he realized that if you were wanting to explain an event in the remote past, you should try to explain it by causes that you see now in operation. And he got this principle from the great geologist Charles Lyell. So, in eastern Washington where I live, there are little patches of white powdery stuff still on the ground from an event that happened in May of 1980. And if you don't know what caused that white powdery stuff, you'd use a standard historical method of reasoning known as the method of multiple competing hypotheses. So, you'd formulate some hypotheses; maybe it was a flood, maybe it was an earthquake, maybe it was a volcanic eruption. Which of those explanations is best according to Darwin and Lyell's principle? Well, it's the volcanic eruption because we have seen volcanoes produce white powdery stuff and floods and earthquakes don't do that. So, if you apply this principle of reasoning, if you look for a cause now in operation and ask yourself what is the cause now an operation that produces digital information? You come to one, and only one type of cause. And that's a mind. Bill Gates says that DNA is like a software program, but much more complex than any we've ever created. What does it take to produce software? It takes a programmer. So, what we think we're seeing with the digital information that's in DNA is not just a problem for Darwinian evolution, but it's a positive indicator of the activity of a mind or an intelligent agent acting in the history and the origin of life."[87]

[87] Meyer, Stephen. "By Design: Behe, Lennox, And Meyer On The Evidence For A Creator -- Uncommon Knowledge with Peter Robinson", Hoover Institution, October 15, 2022, https://www.youtube.com/watch?v=rXexaVsvhCM&t=3s

The brain has been compared to a biological computer. Indeed, computer "thinking" is modeled after human thinking. The type of programming that simulates artificial intelligence is known as neural networking. The human body has been compared to an organic machine, and it is composed of trillions of individual factories called cells. The function of a liver cell is dramatically different from the function of a brain cell, but they contain the same DNA. How does a liver cell know it is a liver cell and is to perform liver functions? The brain cell is programmed to be a brain cell, and liver cells to perform liver functions. It isn't because of luck or by accident. If the unplanned origin of the universe is true, the earliest form of life should have had all the DNA necessary for modern life to exist. Otherwise, we are supposed to believe that complexity accumulates over time, when entropy (decay) is the normal result over time.

Michael Behe explains the conundrum, saying,

> *"The cell is chock full of machines. Machines need many parts. They can't be built by numerous, successive, slight modifications. But let me draw attention to one little sneaky trick that Darwin put into that quotation. He says, "If it could be demonstrated," that something couldn't possibly happen. So, he's putting a burden on his opponents to prove a negative, which science cannot do and never has. No theory has ruled out all rival theories to be accepted. But we have great evidence that it can't, and we have absolutely no evidence that natural selection acting on random mutation could build much of anything."*[88]

Behe astutely pointed out that Darwin employed a clever tactic to neutralize his critics by putting the burden of proof on them

[88] Behe, Michael. "By Design: Behe, Lennox, And Meyer On The Evidence For A Creator -- Uncommon Knowledge with Peter Robinson", Hoover Institution, October 15, 2022, https://www.youtube.com/watch?v=rXexaVsvhCM&t=3s

to prove a negative. Is it possible to "prove" humans did **not** descend from fish? No, but that's asking the wrong question. The correct question to ask is, what is the best evidence that shows a direct relationship between humans and fish because of sexual reproduction over time? We can safely rule out comparative anatomy, one of the three methods used to establish a relationship through evolutionary channels. Fish and humans have no similarities in common. Humans have skin; fish have scales. Humans have lungs; fish have gills. There is no reason to believe or assume any sort of physical relationship between humans and fish—with the possible exception of Neil Shubin's book *Your Inner Fish*, which ironically isn't even about a fish. The book is about an (allegedly) extinct animal named the Tiktaalit—however, I have seen a live creature that looks suspiciously like the fossil swimming in an aquarium. The only physical evidence that a biologist might cite as proof that humans descended from fish is that humans and fish have some genetic sequences in common.

We flunk the comparative anatomy test (that is, unless you've watched *Waterworld* and believe Kevin Costner's gills were real).

Humans and fish share roughly 70 percent of their DNA, which sounds like a lot, doesn't it? Well, consider that humans and apes allegedly share 98 percent of their DNA in common. That's even more impressive...but humans and pigs also share 98 percent of their DNA in common. Humans share about 85 percent of their DNA in common with mice. Yet man's best friend, the dog, only shares around 84 percent of their DNA with humans. Humans have 60 percent of their DNA in common with a banana! Does that mean that humans and bananas are related because of sexual reproduction? In an unplanned universe where Time is the god that makes universal common descent possible, it does.

During his interview on "Uncommon Knowledge", John Lennox makes a significant point:

> *"Now here's where Darwin helped me massively, by expressing, in a letter, a profound doubt. He said, and I'm only paraphrasing cause I haven't got the quote in my head, he said, 'I'm troubled by the fact that if my explanation is correct, then how do we account for the human capacity for rational thought?' He said, 'After all, if we started with lower animals and a monkey's mind,' he said, 'Well, is there any thinking in a monkey's mind?' Now hold that just for a moment because I have lots of fun with my scientific friends. I sometimes ask them, 'What do you do science with?' And of course, they name some expensive machine. And I say, 'No, no.' 'Oh!' they say, 'You mean your...' and they're about to say mind when they realize that's not politically correct, and they say your brain. And I say, 'Okay, I believe the brain and the mind are separate, but what about your brain? Give me the brief history of the brain.' I ask them, and I've done this many times. It's fascinating. And they say, 'Well, the brain in the end is the end product of a mindless, unguided process.' And I smile at them, and I say, 'And you trust it?' I say, 'Now tell me honesty, that computer you use every day, if you knew that it was the end product of a mindless, unguided process, would you trust it?' Now here's the thing, I have spoken with dozens of leading scientists and pushed them on this, and every single one has said no. I said, 'You have a problem. Because you are giving me an argument that undermines rationality.' And they turn to me in and they say, 'Where did you get that argument?' I said, 'Well, firstly from Charles Darwin.' They say, 'I don't believe you.' And then I quote Darwin. Darwin's*

> *doubt. Darwin's doubt about the reliability of human rationality."*[89]

The mind and brain are not the same; they are entirely different things. The human brain is nothing more than physical tissue. The human mind comprises all our thoughts, information, and memories. Evidence suggests the mind can learn new and accurate information while temporarily detached from the brain, a phenomenon called corroborated veridical NDE perceptions. If the mind and brain were the same thing, it would not be possible to corroborate the new information as accurate because it would have been produced by hallucination.

Lennox added, "I'm a mathematician. All mathematicians and scientists are people of faith—not necessarily in God, but they believe in the rational intelligibility of the universe."

Stephen Meyer interjects into the conversation to add, "And the intelligibility of the mind."

Lennox agrees, saying, "Yes. Exactly. And therefore, what do they base that on? If you base that on a mindless, unguided evolutionary process, you're destroying rationality. CS Lewis saw that in the 1940s. He said, 'Any theory that undermines rationality cannot be true because you're using your rationality to get to it.' Alvin Plantinga has worked on it, but the most interesting person who brings it now to the fore is Thomas Nagel, the philosopher in New York. And he says there's something wrong here, because if you follow evolutionary naturalism, it undermines the very rationality you need to believe, not only in evolutionary naturalism, but in any theory at all. So, my major problem, Peter, in all of this, is not the mathematics, that's just an interesting bit of evidence, it's that

---

89 Lennox, John. "By Design: Behe, Lennox, And Meyer On The Evidence For A Creator -- Uncommon Knowledge with Peter Robinson", Hoover Institution, October 15, 2022, https://www.youtube.com/watch?v=rXexaVsvhCM&t=3s

here I am engaged in a rational discipline of mathematics. That all dissolves if the evolutionary naturalistic account is true. In other words, I often say to people, shooting yourself and the foot is painful, but shooting yourself in the brain is fatal."[90]

Thomas Nagel is an atheist, if memory serves correctly. Plantinga is a theist. The perception that a problem exists is not perfectly defined by ideology. Is evolutionary naturalism theoretically possible? Of course, universal common descent must be considered technically possible. However, is it more logical or likely than a common design? Absolutely not. The only way to make evolutionary naturalism more compelling than intelligent design is to dismiss intelligent design as an option without seriously considering it. Dr. Marco Fasoli hammers home the point,

> *"There are fossils of squids that are supposedly hundreds of millions of years old that look indistinguishable from the squids of today. Bugs stay bugs, as David Berlinski famously said. There's no evidence in the past that evolution has happened. Evolution presupposes but cannot account for the origin of life. Scientists have never been able to generate even the very basic constituents of life from simple chemicals despite decades of research and hundreds of millions of dollars of investment. For example, there's no assembly of amino acids which are*

---

90 Lennox, John. "By Design: Behe, Lennox, And Meyer On The Evidence For A Creator -- Uncommon Knowledge with Peter Robinson", Hoover Institution, October 15, 2022, https://www.youtube.com/watch?v=rXexaVsvhCM&t=3s

the constituent blocks of proteins into proteins under abiotic conditions."[91]

Richard Dawkins tells us that we should not trust our own eyes if they see evidence of design in nature. We should trust what the "experts" tell us rather than our own common sense. On the other hand, Buddha told his followers that we should believe nothing that experts tell us unless it agrees with our own reason and common sense. I'm going to have to go with Buddha over Dawkins on this issue.

EVIDENCE FOR THE DIVERSITY OF LIFE

The evidence for the diversity of life is ubiquitous. There are birds, bats, and insects flying in the air. There are mammals, reptiles, amphibians, insects, plants, and trees occupying habitats on land. There are fish and aquatic mammals swimming in the oceans. In addition to a plethora of plants and animals that can be seen by the naked eye, there are also multiple biological kingdoms filled with organisms that are only visible under a microscope, such as bacteria and archaea. I've said this before, and I'll say it again—you don't have to go to church to have a religious experience. Simply watch *Planet Earth* on a high-definition television and marvel at the beauty, mystery, and complexity of creation.

In my previous book, *The God Conclusion*, I mentioned several animals such as the angler fish, which mates in a unique fashion with the smaller male anglerfish physically attaching itself to the female in a parasitic relationship. Once the bond is established between the male and female anglerfish, he loses the ability to survive independently. The platypus is such a unique creature that it is unclear what creature could be called

---

[91] Fasoli, Marco. "7 Scientific Reasons why Darwinian Evolution is a Myth", Radio Immaculata, April 23, 2024 https://www.youtube.com/watch?v=D9zL-f8lSZk

its ancestor—it is a venomous mammal that lays eggs. Anglerfish and platypuses are not the only unique creatures by a long shot. Plants and animals benefit from the existence of each other.

An amazing article on octopi found at a website called Nautilus included this quote speculating about the common ancestry of humans and octopi: "In his book *Other Minds*, the philosopher Peter Godfrey-Smith imagines who this common ancestor might have been. Although we cannot know for sure, it was most likely some kind of small, flat worm, just millimeters long, swimming through the deep, or crawling on the ocean floor. It was probably blind, or light-sensitive in some very basic way. Its nervous system would have been rudimentary: a network of nerves, perhaps clustered into a simple brain. "'What these animals ate, how they lived and reproduced,' he writes, 'all are unknown.' It's hard to imagine something less like us, yet alive, than tiny near-blind worms wriggling on the ocean floor. But we come from them, and so does the octopus."[92]

An opinion stated with such conviction! Why is common descent perceived to be a superior explanation to common design? Do these small, flat worms contain all the DNA necessary to form both humans and octopuses? Wouldn't the flatworms need the DNA to produce both humans and octopuses within their genome, or what other explanation might there be for the introduction of new information? Or do they simply have some DNA sequences in common with both the genomes of humans and octopuses?

---

[92] Bridle, James. "Another Path to Intelligence- Octopus brains are nothing like ours-yet we have much in common.", Nautilus, August 17, 2022, https://nautil.us/another-path-to-intelligence-238534/

Symbiotic Relationships

Symbiotic relationships are close, long-term connections that develop between two species. There are three types of symbiotic relationships—mutualism, where both species benefit, commensalism, where one species benefits and the other has a neutral result, and parasitism, where one species benefits at the expense of the other.

One example of a mutualism relationship is between bees and flowers. Bees benefit from drinking nectar from the flowers, and the flowers benefit from the bees redistributing their pollen.

An (alleged) example of commensalism is the relationship between sharks and remora, also known as suckerfish. The shark provides transportation for the remora, and the remora feeds on the leftovers from the shark's meal but it also feeds on parasites that may attack the shark, providing some benefit to the predator. In reality, because the sharks benefit from the remora's grooming, the relationship could more accurately be described as mutualism.

Tapeworms are an example of a parasitic symbiotic relationship. The tapeworm lives inside the intestines of its host and feeds on nutrients taken from the host. The host derives zero benefit from the presence of the tapeworm and instead loses nutrients to the parasite. Parasites are like leeches—in fact, leeches are parasites. Leeches take blood from their host and give back nothing in return.

Which evolved first—the cotton plant or the cotton rat? Cotton rats get their name from the destructive nature of their relationship to the cotton plant. Even though the rats also eat sugarcane, corn, peanuts, and rice apparently, they do enough damage to cotton plants to merit their unique name. Allegedly, there are fourteen different variations of cotton rats claimed to be individual species, but these species are extremely location specific, such as "found only in western Mexico" or "found only on the Pacific slope of the Sierra de Miahuatlan in the Mexican state of Oaxaca." What if the same creature was found on the Atlantic slope of Sierra de Miahuatlan—would it be identified as a new species? Are any of these rats truly so unique that they should be identified as individual species? Humans do not differentiate into individual species based on location, skin color, hair color, or any other human characteristics. We do not say that pygmies are a subspecies of humans, or that Japanese people are a different species than Americans, even though diet, skin color, and quite a few of the same features are used to differentiate "lesser" creatures such as cotton rats. Why are there allegedly fourteen unique species of cotton rats alone, yet only one species called human beings? Biological classifications are not consistent, and therefore they are mostly useless.

Freshwater Eels

In my opinion, freshwater eels provide some of the best evidence for God's amazing sense of humor, along with the platypus, anglerfish, and octopi. The platypus is amusing because it looks like a weird combination of a bird, a reptile, a mammal, and an amphibian. Scientists in Great Britain were initially shocked to discover platypus specimens were not fakes expertly sewn together. The anglerfish is amusing for the preposterous but curiously effective fishing pole attached to its forehead and for its unique mating habits in which male anglerfish embed themselves into the flesh of the females, becoming parasites. Octopuses are very interesting because they are highly intelligent, as we will learn in a later chapter in more detail.

But freshwater eels also deserve special recognition. A "science educator" calling himself Cole the Science Dude recorded a TikTok video making sensational claims about freshwater eels, and approximately half of those claims appear to be true. For example, Cole claimed that freshwater eels have no reproductive organs, which is true for part of their lives, but scientists have recently observed sex organs in mature eels during the later stages of their life cycles. On the other hand, Cole claimed that freshwater eels can go from freshwater to saltwater and back, and it's true—they are the only animal capable of this feat. Freshwater eels are believed to breed using external fertilization, where millions of sperm fertilize free floating eggs, but this allegedly only takes inside the Bermuda Triangle. Scientists have never found even one single eel egg floating around in the wild.[93]

Like the anglerfish, the mating habits of freshwater eels are unique, to say the least. So, why does science continue to insist a relationship exists between these wildly different creatures based on physical sex? If scientists today cannot say with certainty how these animals reproduce after trying to observe their mating and reproductive habits, how can they predict with any certainty that a direct connection can be formed between these unique animals and humans due to universal common descent? Why should we assume that a common ancestor might be identified through DNA comparisons? Assumptions are inherently dangerous. We can look for sequences in DNA shared by both organisms, but we should not make assumptions as to why those sequences are shared in common or that the only potential explanation can be descent with modifications. If DNA is code and an intelligent agent has the ability to edit or manipulate that code, the exact same effect claimed to be evolution could theoretically occur, but should be called intelligent design because the process can no longer be described as unplanned and undirected.

After his testimony as an expert witness at the Kitzmiller vs. Dover trial (which ruled that teaching intelligent design in the classroom somehow violated the U.S. Constitution), Dr. Ken Miller replied to my email query and described the Robertsonian translocation of human chromosome #2 as

93 Browne, Ed. "Fact Check: Nobody Knows How Eels Reproduce", Newsweek, August 27, 2021, https://www.newsweek.com/fact-check-does-nobody-know-how-eels-reproduce-1623713

creating the impression Scotch tape had been used to attach two chimpanzee chromosomes together. The problem is that Dr. Miller has assumed the evidence only indicates one potential solution, which is (undirected) Darwinian evolution. D. Miller calls the process that created human chromosome #2 "fusion," but insists this fusion takes a long time to occur. Fusion is normally a very fast process that is virtually instantaneous.

One potential explanation is that humans slowly evolved from apes, that is true. However, the process should not be called fusion if significant time is involved. But also, is undirected evolution the only plausible explanation? Obviously not. Humans could have been formed instantaneously by a divine intelligence that used much of the same technology found in the DNA of an ape. Or a series of planned and directed steps undertaken by a source of extreme intelligence also could have applied that supernatural Scotch tape used to connect those two chimp chromosomes into human chromosome #2 if we accept that humans and apes really are that closely related (DNA analysis appears to have established that fact, but there remains some question about the lengths of the different genomes.) Indeed, books have been written with catchy titles such as *99 Percent Ape* that marvel at the fact that the human and chimpanzee genomes share so much information in common.

What we should be marveling at instead is the fact that even a one percent variation in the genetic makeup of an organism can still make a world of difference.

Mass Extinctions in the Fossil Record

The period in the fossil record before complex organisms existed is called the pre-Cambrian. One of the most important events documented in the fossil record is known as the Cambrian Explosion, an event during which every conceivable body plan known to exist today, plus several others that became extinct, all appeared in the fossil record within a very short timeframe, geologically speaking. If evolution allegedly happens gradually, over exceedingly long periods of time, how do we explain the sudden appearance of every known body type in the fossil record, all beginning to exist at approximately the same time? Estimates suggest that roughly half of the living organisms populating the Earth during the Cambrian period were exterminated during the Cambrian extinction event. Estimates also suggest that up to 95 percent of all life on Earth was wiped out during the Permian extinction, which took place only 253 million years ago. There have been at least two mass extinction events since the Permian extinction. The Triassic-Jurassic extinction event allegedly occurred approximately 200 million years ago and wiped out all the dinosaurs, followed by the most recent Cretaceous-Paleogene extinction event roughly 65 million years ago.

According to paleontologists, there have been at least five mass extinction events in Earth's history.[94] The largest mass extinction events, the approximate percent of life killed, and how many million years ago the event took place were:

Event Name	Percent Killed	Millions of Years Ago
Ordovician/Silurian	85	440
Devonian	75	365
Permian	95	253
Triassic/Jurassic	80	201
K-T	75	66

Paleontologist Michael Benton wrote in his book *When Time Nearly Died*,

[94] Dutfield, Scott. "The 5 mass extinction events that shaped the history of Earth—and the 6th that's happening now." LiveScience, May 17, 2021, https://www.livescience.com/mass-extinction-events-that-shaped-Earth.html

> *"When I was a student of paleontology in the late 1970s, I remember being puzzled by some extraordinary differences of view. These were not disagreements—that sort of thing is common enough, and such disputes are usually resolved eventually by the discovery of some new evidence that settles the matter one way or another. This dispute was over the greatest mass extinction of all time: some paleontologists accepted that, around 250 million years ago, all of life on Earth came very close to extinction, while another group argued that nothing at all really happened. As I delved deeper into the question, I was convinced that there had indeed been a mass extinction of huge magnitude about 250 million years ago, at the end of the Permian period. But why had some paleontologists denied it? The extinction deniers were not creationists, flat-earthers, or members of some other fringe group. They knew their fossils, and yet their reading of the fossil record seemed to tell them that the end of the Permian passed with only the merest blip, the smallest disturbance, and really it was nothing to be concerned about."*[95]

It's a fascinating point Benton has fixated upon because he had just subtly shown his disdain for "creationists" by lumping them in with "flat-earthers" and other fringe groups. Instead, it was a group of respected colleagues who had looked at the exact same evidence as him and came to a different interpretation of what that evidence meant. He's astonished that anyone with any credibility could look at the same evidence and reach a different conclusion than he did.

Benton was talking about the Permian extinction, which allegedly wiped out more than 90 percent of existing terrestrial

---

95 Benton, Michael. *When Life Nearly Died: The Greatest Mass Extinction of all Time*. Page 7. New York. Thames and Hudson. 2003. Print.

species and many aquatic species, leaving a so-called "lucky" survivor called Lystrosaurus to become the primary ancestor of the dinosaurs. He wrote,

> *"I tried to explain that [Lystrosaurus] was in fact a very ordinary animal. It had no special survival qualities that other animals lacked. It was simply lucky. 'Why?' they kept asking, 'did Lystrosaurus survive, and nothing else?' A clear adaptive statement was required. I explained that* Lystrosaurus *was a survivor of a sort, but it was not particularly fast, fearsome, or intelligent. It was really like something of a Triassic pig in appearance, and perhaps even in habits, since it was probably a generalist, without specific adaptations in its diet, living requirements, or mode of locomotion. The point is that good fortune is a characteristic of mass extinctions. The survivors are more lucky than specifically adapted."*[96]

The words in the quote implying that luck influences our existence in an unplanned universe have been emphasized. There either is a plan, or there isn't. But if there really isn't a plan, then anything resembling order must only be an illusion. I had been told the theory of evolution had nothing to do with luck, but here Benton is saying that luck was crucial to the survival of Lystrosaurus. Is luck involved, or not?

Neil deGrasse Tyson asserts,

> *"There was a while there where we looked at the extinction records of species on Earth and found some period that repeated where the fossil record showed a dramatic drop in the species count from one layer to the next in the geological sediment, and there is nothing in the solar system that has a twenty-million-*

[96] Ibid. Page 22.

> *year rhythm. So, someone suggested maybe it's in a binary star system where there's another star that plunges in through the solar system, coming through the Kuiper belt, and then goes back out in this dance with the sun. So, we wouldn't have seen it in our civilization, but when it does that, it disrupts the Kuiper belt gravitationally and if you do that, you will send a rain of comets down into the inner solar system and then you could render many life forms extinct on Earth, just the way we lost the dinosaurs from an asteroid. And they even gave a name to it—they called it Nemesis. But we took a closer look at the data, and it turned out it had been filtered in a way that revealed rhythms that were not really there. So, that concept has evaporated, but it got people going for a while. It got a lot of press attention."*[97]

Why yes, getting press attention by making incredible claims is far more important than getting it right these days, isn't it? It is an interesting mystery why the fossil record appears to show a pattern or cycles of explosive new life forms, a period of stasis, followed by a mass extinction.

Benton also wrote,

> *"Scientists are evidently human too. They can be prejudiced, they can be scared, or constrained. Catastrophic extinctions in the geological past is a beautiful example of an idea that was presented in the 1820s, that was firmly crushed by Lyell in the 1830s, and could only raise its dangerous head again in the 1980s. It took 150 years for geologists to dare accept the obvious, that there truly have been mass*

[97] Tyson, Neil deGrasse. "Nemesis Concept with Neil deGrasse Tyson." UniverseLair, May 24, 2025. https://www.youtube.com/shorts/ugrPkgI1xdY

> *extinctions in the past and that structures on the Earth's surface that look like impact craters actually are impact craters."*[98]

He added these words of wisdom: "As one grows older, one realizes how little one knows: 'the more you learn, the more ignorant you become.' The joy of being a scientist is to discover this."

I would call that the joy of being a human being. Are all human beings scientists? More importantly, are not all scientists human beings?

Examples of Intelligent Design

In his book *The Way of the Cell*, chemistry professor Franklin Harold wrote: "We should reject, as a matter of principle, the substitution of intelligent design for the dialogue of chance and necessity, but we must concede that there are presently no detailed Darwinian accounts of the evolution of any biochemical system, only a variety of wishful speculations."[99]

Why should we reject intelligent design if Darwinian ideology fails to explain the complexity of life? What principle is involved in rejecting a potential explanation a priori, without a thorough examination?

---

98 Benton, Michael. *When Life Nearly Died: The Greatest Mass Extinction of all Time.* Page 304. New York. Thames and Hudson. 2003. Print.

99 Harold, Franklin. *The Way of the Cell: Molecules, Organisms, and the Order of Life.* Page 235. New York. Oxford University Press. 2001. Print.

Echolocation Navigation

In the foreword, I promised to provide some evidence of intelligent design.

Humans have developed a sophisticated technology called sonar. Soundwaves are used through water to navigate a vessel, measure distances and identify objects both on the surface and underwater, and to communicate with other vessels. Sonar is a credit to human ingenuity and quite obviously the product of intelligent design. Sonar is based on echolocation navigation, a natural ability found in bats, dolphins, and whales.

Sonar	**Echolocation**
Uses man-made equipment and technology	Natural ability of some animals
Sound waves transmitted artificially from a device	Sound waves are produced by the animal's body
Works in air, water, or other media	Typically functions only in air or water
Long range detection, up to several miles	Shorter range detection, hundreds of feet or less
Lower frequency, longer wavelength sound waves	Higher frequency, shorter wavelength sound waves
Used for navigation, detection, mapping	Used for navigation, prey detection, object avoidance

Data recorded and analyzed extensively	Real-time processing only by the animal's brain[100]

From where did humans get the idea for sonar? They got the inspiration by observing how bats, whales, and dolphins use their natural abilities of echolocation for navigation. These inventors copied and mimicked what they had learned through observation. Humans deserve the credit for inventing sonar, but no credit for the concept of echolocation navigation. Naturally (pun intended), echolocation is superior in the sense that the animal can process the information in real time in its brain, while human-developed sonar must be analyzed and interpreted, but there is a tradeoff because sonar can work at much greater distances.

The fact that humans can borrow an idea from Nature and create a technology that provides an obvious benefit from its use is a clear example of intelligent design, so sonar is a product of intelligent design. However, the product is copied from an example that modern biologists would claim resulted from a virtually endless number of mutations and adaptations over a prolonged period of time that happened for no reason, which seems to be a silly argument to me. Clearly, there is a purpose for echolocation navigation. Whales, bats, and dolphins use this ability to find food and avoid dangerous obstacles in the water. Without this ability, these creatures would starve.

Humans needed the ability to avoid dangerous obstacles in the oceans, so they stole the idea and developed the technology to achieve that same objective that some animals can easily accomplish using only their God-given abilities.

---

100 Brodwin, Maya. "What is the difference between sonar and echolocation?", Birdful, February 4, 2024. https://www.birdful.org/what-is-the-difference-between-sonar-and-echolocation/

ANALYSIS OF THE DIVERSITY OF LIFE

My objection to Darwinian evolution does not result from a personal animus for the idea. My objection is to the idea that the process is unplanned and undirected. I've devoted considerable study to developing what I call my biology thought experiment, an attempt to understand how any existing organism could change enough through descent with modifications to the point where offspring could be classified as a new or different type of organism.

My Biology Thought Experiment

Problem: How do we get from Old World Ape to Australopithecines, or whatever the first ancestor was, from a true primate ancestor, and to a human being?

Potential Answer: According to Darwin's theory, macroevolution takes place as a series of incremental steps that occur over a prolonged period of time. How do we get to a population of mutated Old-World Apes that now possess 46 chromosomes and Human Chromosome #2 (HC2) from an existing population that remains apes that have 48 chromosomes, smaller craniums, and fur? It is an interesting and particularly challenging question.

According to biologist Jerry Coyne, isolation is a key factor in speciation. One breeding population splits into two and becomes separated from each other to prevent interbreeding. Most apes have 48 chromosomes, whereas most humans have 46 chromosomes. In some rare cases, humans have been documented as having different chromosome counts, but even so, a person with 44 chromosomes is not short two chromosomes worth of DNA in their genome—their genetic information is simply organized differently than most human beings. The information is all still there, or the subject would not be an intact person.

That first mutation could logically be any of a number of morphological changes that must occur for an ape to gradually become a human—loss of body fur, changes in cranium size, skeletal changes, etc. However, choosing the formation of human chromosome #2 by the fusion of two ancestral ape chromosomes as the mutation to be studied has benefits. This particular mutation could be considered neutral or perhaps even beneficial, unlike the loss of body fur, loss of muscle strength, or other deleterious but necessary mutations that must occur for an ape to shape-shift into a human being. Those mutations would be more noticeable than a genetic change, and we have observed how nature tends to deal with abnormal-looking offspring. They are typically shunned and sometimes killed by their own parents. This would be important because the population at large would not be able to notice any sort of morphological differences that might lead to discrimination by the majority of the breeding population, which rarely or never ends well. We begin with population Zero (P0), which is consisted only of apes with 48 chromosomes. A second breeding population, Population One or P1, becomes physically separated from population P0, and over time, mutations begin to randomly occur. At first, they are extremely rare. Because the differences between these members of the same breeding population are very limited to the single change, we should assume these mutated offspring would be able to mate with other members of the population and still be able to produce fertile offspring. However, we should assume the dominant genetic trait would be for offspring to have 48 chromosomes, not 46.

The problem is that as long as there are any members of P1 with 48 chromosomes, that will remain the dominant genetic trait. At best, members of the population with HC2 might "evolve" from a rare mutation to a recessive gene, but the only way you're going to get a population exclusively formed of members with HC2 instead of 48 chromosomes is by killing off every member of the group with 48 chromosomes, because that is the dominant trait. If there is any sort of breeding cross-contamination with P0 at any point in time, then any evolutionary progress would immediately be wiped out in P1. As long as P1 has members with 48 chromosomes, that will remain the dominant characteristic.

We can imagine how HC2 could begin to spontaneously pop up as a rare mutation in our hypothetical scenario. We can even see the possibility that the rare mutation might become more common over time, eventually becoming a recessive trait characteristic in the gene pool. As we've learned from so-called genetic experts: "The final step would be a situation where those with 46 chromosomes became isolated from those with 48 and where the 46'ers did better for some reason than the 48'ers."[101]

That sounds perfect in theory, but how does it work in reality? It's a little too perfect. Remember, in our little thought experiment, the members of the population with only 46 chromosomes wouldn't be able to detect any differences between themselves and members of the population with 48 chromosomes. So, how does the group with only 46 chromosomes become perfectly isolated from the primary group? And why does it happen? That is the crucial question — how indeed does breeding population P1 become exclusively populated by apes with only 46 chromosomes, one of which is now HC2? Not meaning to sound trite, but this is starting to sound like an act of God would be required for this to happen, but then we would no longer be talking about evolution, but intelligent design.

There would seem to be only a few ways for the entire population to be exclusively populated by members with only 46 chromosomes: every member of the population with 48 chromosomes either dies or becomes celibate. For as long as the

---

[101] Starr, D. Barry. "How Did Humans Go From 48 to 46 Chromosomes?", The Tech Interactive, January 16, 2013, https://www.thetech.org/ask-a-geneticist/48-46

population contains any sexually active members with 48 chromosomes, that will be the dominant characteristic of any offspring produced. In our hypothetical thought experiment, macroevolution fails to produce a new species because dominant genes will always dominate, and recessive characteristics will always remain secondary options within the constraints of the existing genome.

According to biologist Jerry Coyne, the key factor in creating a new species is the isolation of an existing breeding population from within a larger breeding herd. However, when we observe the reality of our natural world, we do not see evidence of new species forming from existing species, either. The cichlids of Lake Victoria became isolated and interbred for many generations, creating a diverse variety over time. However, the cichlids remained cichlids. They never evolved into a new type of fish. Of course, the ichthyologists have claimed that up to 600 new species of cichlids were identified in the lake, but really, they were different varieties of the same basic type of fish, which remains the cichlid. The fish underwent no new morphological changes.

At the conclusion of our thought experiment, logic tells us that a new species cannot be created by biological evolution because even if an existing population is split and becomes separated into two different breeding populations, dominant characteristics will always remain dominant. Recessive traits don't become dominant in an organism over time, much less exclusive to the new population. The North American black bear and the Asiatic black bear are basically the same bear, except the Asiatic version usually has a blaze marking on its chest. The North American black bear typically does not have the blaze, but some do. Some North American black bears have a blazed chest, while some Asiatic black bears don't have blazes. The point is, they don't really seem to be two unique species of bear, but merely two varieties of the same bear species that live on different continents. One variety tends to have chest markings, and the other does not. Otherwise, it's the exact same bear.

For humans to have evolved from apes, there needs to be a first step in the process. For the sake of argument, in this thought experiment, we have assumed that HC2 was the first mutation and tried to understand how that mutation could become the dominant feature of an entire breeding population, and the short answer is, it simply doesn't seem to be possible. **There simply doesn't seem to be a logical way for that separated population to develop that initially rare characteristic into a dominant trait and finally an exclusive trait, no matter how much time elapses**, unless we assume the laws governing sexual reproduction might have been variable over the course of time.

However, then we would be forced to contemplate a world in which the laws that govern our existence have not always been consistent, so there would be no logical way to predict future events based on the past. Even if that model is true, it would prove any effort to understand our world by applying the scientific method would ultimately prove futile. My goal in this thought experiment has been to combine my limited knowledge of genetics with my limited understanding of macroevolution theory to produce a narrative by which a single step in the evolutionary process might be achieved, but elementary logic has unfortunately prevented the success of my effort. It is far easier merely to suggest that one group splintered from the other and simply "did better for some reason" than it is to contemplate and ultimately explain how it might have possibly happened.

Therefore, the result of my attempt to logically construct a plausible scenario in which macroevolution truly might occur has failed miserably, because I couldn't envision a logical way to even take the first step. If macroevolution is true, bears and elephants would share a common ancestor due to generations of sexual reproduction. Naturally, apes and humans would also share a common ancestor. We should also expect to see intermediate creatures that bridge more gaps between organisms. In essence, every living organism on Earth would share a common ancestor with every other organism, which is a principle known as universal common descent. Ultimately, all because of sexual reproduction. The problem is, I simply cannot conceive of any logical means by which even that first step in the process might be achieved, unless we change the rules in the past and allow them to be different from rules in the present. But if the past is inconsistent, then our predictions become useless, nothing more than wild guesses.

Time and Luck, the Secular Gods of Methodological Naturalism

For a secular worldview to succeed, certain ideas or concepts must be treated with the same sort of reverence a religious person might feel toward a deity. I like to say Time can be said to be the secular god of Methodological Naturalism because Time is the hero of the plot that attempts to explain how life initially came to exist. Richard Dawkins says that it is only because I cannot imagine and thus underestimate the vast amount of time necessary that I could reject the idea of common descent. However, I do understand the concept of time fairly well. I also understand how mass extinctions would considerably shorten the amount of time available for mutations to accumulate and become improvements—and this is assuming the current dating methods are trustworthy and reliable, though we have reason to believe that may not be true. Rocks produced by the Mount Saint Helens volcanic eruption only a few decades ago were dated as being several hundred thousand years old.

Once again, there are two basic worldviews that argue the origin of the universe (and the origin of life) were either planned or unplanned. A planned universe could be any age. It could only be thousands of years old, but created to appear billions of years old. Or it could literally be billions of years old, as scientific evidence suggests. The only reason to believe the universe might only be thousands of years old is if we stick with a strict, literal interpretation of the Bible. Based on the timeline produced by Bishop Usher, the age of the Earth has been estimated to only be around six thousand (6,000) years old. Old Earth Creationists and intelligent design advocates do not adhere to the Young Earth Creationist timeline because the Bible makes it clear God does not experience time the same way humans do, which makes sense. After all, God is an eternal being without beginning or end. According to the Bible, a day is like a thousand years to God.

Paley's Watchmaker Analogy

William Paley famously proposed the Watchmaker argument, which says that a man walking through woods might accidentally kick a stone with his foot and imagine that the stone had always been there. He then substitutes a watch for the stone and recognizes the watch is a product of intelligent design, which implies the existence of a designer. Attempts to rebut the Paley analogy focused on the need for God to be more complex than His creation, which seems to be a logical conclusion. Richard Dawkins even wrote a book titled *The Blind Watchmaker*. There is a problem with Paley's Watchmaker, but it isn't the problem that Richard Dawkins suggested. The problem with Paley's Watchmaker analogy is that Paley assumed the universe itself was eternal, and the rock could have literally been there forever. We know the universe had an origin, or at minimum, a starting point before which there were no stars or planets. Or rocks.

Planned Versus Unplanned

It can be ascertained through logic that unplanned modifications and improvements to existing life forms are absurdly improbable. This is due to the extremely improbable origin of the universe, the extremely improbable expansion of the universe, and the extremely improbable origin of life all of which are necessary for life to exist and diversify. The improbabilities compound as each future event requires the success of one or more previous events. Simply stated, there is no good reason to believe that sufficient Time allows algae to turn into fish that turned into apes that ultimately turned into humans. This is granting the awesome power of a god to Time itself. However, entropy is the enemy of Time, and Time is only a measurement with no creative power.

Assuming we have a scale where zero percent represented certainty that the universe was unplanned and one hundred percent represented absolute certainty of a planned universe, we should begin with a neutral (fifty/fifty) position because the universe could conceivably be either planned or unplanned. We don't know because we haven't looked at any of the scientific evidence yet. We need evidence to move the needle in one direction or the other. The evidence for cosmic fine-tuning immediately makes the origin of the universe a highly improbable event. This improbability associated with the unplanned universe is far more than fifty percent in favor of the planned universe because of multiverse and other "solutions" offered to address the fine-tuning problem don't address the problem of First Cause. The precision of cosmic inflation further pushes the needle in the direction of the planned universe...even though we can never achieve one hundred percent absolute certainty of a planned universe, as we examine each of the miracles necessary for existence, we see the odds that the universe was planned improve, and conversely, the odds of an unplanned universe continue to worsen.

The so-called experts in biology will admit that living organisms appear to have been designed, but then they insist that appearances can be deceiving. Biology professors can also be deceiving. They pretend the idea of intelligent design is inconceivable even though the evidence for design is overwhelming. Humans steal the best ideas from nature all the time, and the result is typically called the product of intelligent design. How can an unplanned model become the basis of a planned and intelligent (but inferior) copy? The human body is an amazing organism with multiple complex systems essential for the body's survival—the skeletal system, cardiovascular system, and immune systems to name just a few of the vital systems in the human body. In the final section, we will explore the complexity of the design of a human being. The human body contains systems within systems that are essential to its function. It is extraordinarily difficult, if not impossible, to believe they exist due to random chance.

Another organism with features that smack of intentional design is the anglerfish, which has a lure attached to its forehead and a unique mating habit that doesn't seem to be shared by other organisms—the smaller male anglerfish literally embeds itself into the flesh of the larger female in parasitic fashion. The only other creatures that apparently share this unique method of mating are other "species" of anglerfish. And of course we have the platypus, an animal so unique that taxidermists in England believed the first specimens sent from Australia had been faked. And let's not forget the mysterious and unique mating habits of freshwater eels. Why should we believe all these creatures share a common ancestor, not an uncommon designer?

People tend to assume that my criticisms of Darwinism are due to my personal religious beliefs, but that is simply untrue. I could believe that evolution was true if it weren't for the planned versus unplanned universe argument. I know of several famous scientists who believe in theistic evolution, namely Drs. Francis Collins and Ken Miller. It is for purely logical reasons alone that I must continue to object to the theory of evolution as a viable explanation for the diversity of life we observe on Earth. God would have to be deistic for evolution to be true. If the origin and rapid expansion of the universe were both unplanned and the origin of life was unplanned, then it stands to reason that the diversification of life would also be unplanned. But another word for unplanned is "lucky", and nobody believes in the kind of luck that would be necessary for luck to explain the power and intelligence of a creator God.

Darwinian evolution is often treated with the same sort of reverence one might expect to find in a religion. Richard Dawkins even said that to deny the evidence for evolution was like trying to deny the evidence for the Holocaust. The problem with that comparison is that we do have eyewitness testimony to the Holocaust, but macroevolution has never been and cannot be observed. It can only be inferred after the fact, and the problem with that is we have examples of theories being proved wrong by observation. Sadly, in spite of the overwhelming evidence for the Holocaust, there are people even today who deny it happened, or insist it wasn't as bad as people make it out to be.

For example, in my book *The God Conclusion*, I wrote about the sailing stones of Racetrack Playa as an example of a preferred and widely accepted hypothesis being disproved through observation. There were only two potential explanations for the movement of these large rocks on the remote desert plain known as Racetrack Playa: wind or ice. They had a fifty/fifty chance of a correct guess. Wind was assumed to be the culprit after an experiment seemed to rule out ice as an explanation, but observation ultimately showed that ice was responsible for the movement of the rocks.

If we believe in universal common descent with enough conviction, we might convince ourselves the evidence for evolution is so overwhelming that the amount of luck necessary for the success of the miracles needed to precede it is irrelevant and doesn't matter. If evolution has become your religion, nothing else does matter.

Symbiosis

The concept that makes unplanned evolution so difficult to believe is symbiosis. Probably the best example of symbiosis is the sea anemone and the clownfish. The clownfish could not survive for long in the ocean without the benefit of protection from the sea anemone that violently stings any predators that would otherwise eat the clownfish, but to which the clownfish apparently has immunity. To reciprocate for that protection, the clownfish eats small invertebrates that would otherwise harm the sea anemone, and the fecal matter of the clownfish provides nutrients to the plant. What is the likelihood that two vastly different organisms would evolve separately and yet learn to cooperate in a mutually beneficial manner?

These harmonious relationships create a delicate balance in nature.

Food Chains

Naturally, scientists can't be satisfied calling a food chain a food chain—they must call it a "trophic cascade" to make it sound more science-y. That doesn't change the fact that at the top of the food chain are predators, and at the bottom are prey. Predators help control and manage the population of the prey by eating the oldest, weakest, and dumbest members of the herd.

One of the most remarkable videos I can recall ever seeing was titled "How Wolves Change Rivers"[102]. The short video documents the effects of reintroducing wolves into Yellowstone Park after the wolves had been hunted to extinction 70 years earlier. As the video's title illustrates, the wolves not only affected the ecosystem of the national park, but they also caused the physical geography to change as well. It's a truly remarkable video. The hyperlink can be found in the index. If organisms did not exist for some purpose or reason, food chains would not exist. Introducing the wolves back into Yellowstone would have only produced more violence, more predatory behavior, and

[102] Monbiot, George. "How Wolves Change Rivers", Sustainable Human, February 13, 2014, https://www.youtube.com/watch?v=ysa5OBhXz-Q

more death. Instead, harmony in nature was restored, and the ultimate effect was nothing short of remarkable.

CONCLUSION

In an article about the famous Wistar symposium, the author wrote, "The mathematician D.S. Ulam argued that it was highly improbable that the eye could have evolved by the accumulation of small mutations, because the number of mutations would have to be so large and the time available was not nearly long enough for them to appear. Sir Peter Medawar and C.H. Waddington responded that Ulam was doing his science backwards; the fact was that the eye had evolved, and therefore the mathematical difficulties must only be apparent. Ernst Mayer [SIC] observed that Ulam's calculations were based on assumptions that might be unfounded and concluded that 'Somehow or another, by adjusting these figures we will come out all right. We are comforted by the fact that evolution has occurred.'"[103]

"Comforted" by the fact? Not confident in the fact? Why did these scientists need for the theory of evolution to be true? Were they that afraid of an all-powerful God?

Everyone knows that the absolute best science is done by reaching a conclusion first and then fitting the facts to the narrative that supports it. (That was sarcasm, for the record.) No. That isn't science at all. That's a religious belief.

Speaking of religious beliefs, Genesis 1:11-30 reads:

> *Then God said, "Let the land produce vegetation: seed-bearing plants and trees on*

[103] Klinghoffer, David. "Wistar: Been There, Done That", Evolution News & Science Today, November 8, 2018, https://evolutionnews.org/2018/11/wistar-been-there-done-that/

*the land that bear fruit with seed in it,
according to their various kinds." And it was
so.* 12 *The land produced vegetation: plants
bearing seed according to their kinds and trees
bearing fruit with seed in it according to their
kinds. And God saw that it was good.* 13 *And
there was evening, and there was morning—
the third day.*

14 *And God said, "Let there be lights in the
vault of the sky to separate the day from the
night, and let them serve as signs to mark
sacred times, and days and years,* 15 *and let
them be lights in the vault of the sky to give
light on the earth." And it was so.* 16 *God made
two great lights—the greater light to govern
the day and the lesser light to govern the night.
He also made the stars.* 17 *God set them in the
vault of the sky to give light on the earth,* 18 *to
govern the day and the night, and to separate
light from darkness. And God saw that it was
good.* 19 *And there was evening, and there was
morning—the fourth day.*

20 *And God said, "Let the water teem with
living creatures, and let birds fly above the
earth across the vault of the sky."* 21 *So God
created the great creatures of the sea and every
living thing with which the water teems, and
that moves about in it, according to their kinds,
and every winged bird according to its kind.
And God saw that it was good.* 22 *God blessed
them and said, "Be fruitful and increase in*

number and fill the waters in the seas, and let the birds increase on the earth." [23]

And there was evening, and there was morning—the fifth day. And God said, "Let the water teem with living creatures, and let birds fly above the earth across the earth across the vault of the sky." So, God created the great creatures of the sea and every living thing with which the water teems, and that moves about in it, according to their kinds, and every winged bird accordin to its kind. And God saw that it was good.

God made the wild animals according to their kinds, the livestock according to their kinds, and all the creatures that move along the ground according to their kinds. And God saw that it was good.

[26] *Then God said, "Let us make mankind in our image, in our likeness, so that they may rule over the fish in the sea and the birds in the sky, over the livestock and all the wild animals,*[a] *and over all the creatures that move along the ground."*

[27] *So God created mankind in his own image, in the image of God he created them; male and female he created them.*

[28] *God blessed them and said to them, "Be fruitful and increase in number; fill the earth and subdue it. Rule over the fish in the*

sea and the birds in the sky and over every living creature that moves on the ground."

[29] Then God said, "I give you every seed-bearing plant on the face of the whole earth and every tree that has fruit with seed in it. They will be yours for food. [30] And to all the beasts of the earth and all the birds in the sky and all the creatures that move along the ground—everything that has the breath of life in it—I give every green plant for food." And it was so.

God wasn't done with his creation. He also created the science of taxonomy, according to the second chapter in Genesis: [19] Now the Lord God had formed out of the ground all the wild animals and all the birds in the sky. He brought them to the man to see what he would name them; and whatever the man called each living creature, that was its name. [20] So the man gave names to all the livestock, the birds in the sky, and all the wild animals.

The theory of evolution allegedly describes an unplanned process. The Bible describes a planned process. It gives us an order in which creation occurred. Theistic evolution is the ultimately untenable attempt at compromise between planned and unplanned solutions by adding God to evolution. Only creation requires a Creator. Intelligent design does not necessarily require a God, but it does require an unidentified source of intelligence and gives that source the powers of a god.

Creationism fits neatly within the planned universe paradigm, if you don't get hung up on requiring the age of the Earth to be only six thousand years old when it's probably much older.

MIRACLE 5
THE ORIGIN OF INTELLIGENCE

INTRODUCTION

Intelligence may be defined as a measure of the ability to acquire and apply knowledge and skills to solve problems. Intelligence has also been defined as the ability to acquire, understand, and use knowledge. Is a bird intelligent? Not long ago, a cardinal got trapped inside my garage. The garage door was open, but the bird had flown too close to the ceiling and got trapped between the open door and the sheetrock on the garage ceiling. The poor bird kept flying into the ceiling and hitting its head against the sheetrock. Finally, I closed the garage door. The bird flew lower and landed on a board only two or three feet off the ground. I opened the garage door and used a stick to keep the bird from flying back up to the ceiling. Instead, he flew outward and into the safety of the night. Had I simply closed the garage door and trapped the bird inside, he likely would have died.

Compared to a human being, the bird was not displaying an analogous degree of intelligence. Indeed, this is probably why humans sometimes describe those humans exhibiting considerably less intelligence than themselves as being "bird-brained." When the problem of the cardinal trapped inside the open garage initially presented itself, the bird was too close to the ceiling. When the bird looked outward from that vantage point, he only saw walls and the rolled-up metal door blocking his exit. He simply lacked the intelligence to realize a slight change in perspective would solve his problem, so the poor bird kept beating his head against the ceiling. All I had to do was get the cardinal closer to the ground and looking toward open space to convince him to safely leave the unnatural environment inside the garage, where we both knew he would die if he didn't leave. The brain weight of an average adult human is three pounds, or approximately 1250 grams. The average brain weight of a cardinal is only about 1.6 grams. Based on raw weight differences alone, the human brain is smarter than a cardinal's brain. However, if we go by brain-to-body-mass ratios, the average human has a 1:52 ratio, and a cardinal has a brain-to-body-mass ratio of 1:28, which means a cardinal's brain actually has a larger percentage of total body weight than a human's brain. The bird might not have been smart enough to fly lower without encouragement, but he was smart enough to quickly fly to safety when given the opportunity. His survival instinct was working quite well.

For most of this book, we've described the ultimate argument as being whether the universe might be planned or unplanned. It is a binary alternative, meaning there are only two options from which to choose. There is another binary alternative when looking at the construction of the universe: top-down versus bottom-up. A planned universe is created from the top down. The best explanation for the origin of intelligence is the existence of a higher intelligence. In an unplanned universe, life is created from the bottom up. Intelligence is (allegedly) created by non-intelligent physical matter, perhaps as a chemical reaction. If that is true, as John Lennox asks, how can we trust our own intelligence if it originates from unintelligent, blind processes?

Lucy

Lucy is a science fiction movie starring Scarlett Johansson. The film explores the limits and capabilities of human intelligence in a thrilling scenario.

Lucy (Scarlett Johansson) has accidentally ingested a lethal amount of a synthetic drug that exponentially increases the brain's ability to learn new information. Realizing that the overdose will kill her and she won't survive much longer, Lucy reaches out to Professor Samuel Norman (Morgan Freeman), an expert on intelligence and human cognition. Norman is intrigued by Lucy's intellect as she explains her predicament. He asks, "But if humans are not the unit of measure and the world isn't governed by mathematical laws, what governs all that?" Her response is both stunning and fascinating: "Film a car speeding down the road. Speed up the picture infinitely, and the car disappears. So, what proof do we have of its existence? Time gives legitimacy to its existence. Time is the only true unit of measure that gives proof to the existence of matter. Without time, we don't exist."[104]

Clever dialogue. In an unplanned universe with billions of years of time that can make virtually anything eventually possible, Time can have the creative power of a god but not the intelligence of one. Even so, we have no actual evidence that Time possesses any sort of creative abilities. Time also gives us entropy, which is the ability to destroy. But this is philosophy coming from a Hollywood movie. What exactly did we expect from a business that routinely asks us to willingly suspend our disbelief before we buy a ticket? Did we expect the story to make perfect sense? The audience isn't looking for a realistic plot; they are happy if the stars look pretty, the action looks believable, and the story is even remotely plausible. If not, they will take wildly entertaining as a substitute for believable stories. What the audience does not want are stupid, boring, or unrealistic stories. They want to be entertained.

Lucy is not realistic, but neither is it stupid or boring. It isn't plausible, but it doesn't really need to be because it is very

---

[104] Johannson, Scarlett. "Time is the ONLY True Measure! — Movie Flash Cuts, Lucy's Mind-Blowing Revelation (Lucy 2014)", May 21, 2025, https://www.youtube.com/shorts/7IVLnkXpej4

entertaining. If you stop to think about the plot for a minute, it becomes obvious the story is ludicrous. So, don't stop and think.

For example, at the beginning of the story, Lucy was a drug mule, which explains how she got the drug into her body...an illegal drug that makes you smarter instead of getting you high or addicted. Why would a drug that increases intelligence need to be sold on the black market? If you stop to think, it will ruin the movie. Don't take the movie seriously, and you'll enjoy it a lot more.

Interestingly, the Bible doesn't really talk much about the intelligence of living creatures. Instead, the Bible speaks of knowledge and wisdom, and of the ability to discern what is right and true.

THE EVIDENCE FOR INTELLIGENCE

In 1975, Dr. Paul Cano of the National Centre for Scientific Research (CNRS) discovered that oak trees under duress from attack by caterpillars reacted by increasing the quantity of tannin and phenol in their leaves as a defense mechanism. The strategy inhibited the growth of caterpillar larvae and saved the trees from annihilation.

According to this news article from New Scientist[105], dated fifteen years after the discovery by Cano, another mystery was solved after three thousand South African antelope called kudu suddenly died for unknown reasons. Giraffes wandering freely around the game ranches also ate leaves from the same acacia trees, but only the kudu were dying, and by the thousands. Zoologist Wouter Van Hoven eventually discovered that the

[105] Hughes, Sylvia. "Antelope activate the acacia's alarm system." NewScientist, September 29, 1990, https://www.newscientist.com/article/12717361-200-antelope-activate-the-acacias-alarm-system/

downwind acacia trees were producing excess tannin to toxic levels in their leaves because trees upwind had emitted ethylene to warn them of the danger. In other words, the acacia bushes had their own alarm system that the kudu's overgrazing triggered, and their self-defense mechanisms killed the kudu.

Neil DeGrasse Tyson said,

> *"Who are we to say that we're intelligent? I pose that not as a joke, but as a very serious question. We define ourselves to be intelligent in ways that no other creature can rival. What do we credit that intelligence to? So, you look at the genome, and let's take the chimp—and we have, what is it? High 90s percent identical, indistinguishable DNA. And the chimp does not build the Hubble telescope. So, we must then declare that everything we say about us that is intelligent is found in that 1.5 percent difference. Let me invert that question—if the genetic difference between humans and chimps is that small, maybe the difference in our intelligence is also that small. Maybe the difference between stacking boxes and reaching a banana, whatever are these rudimentary things a chimp does that the primatologists roll them forward and boast about, which of course our toddlers can do…maybe the difference between that and the Hubble telescope is as small as that difference in DNA."*[106]

If science tells us through observation that even plants without brains are intelligent, why shouldn't we believe that humans with complex brains, the ability to make tools and machines, and the ability to communicate with symbols and writing are not intelligent? There is no reason to believe Neil deGrasse

---

[106] Tyson, Neil deGrasse. "We're intelligent? w/Neil DeGrasse Tyson.", Universe Lair, February 25, 2025, https://www.youtube.com/shorts/LSXlh_XVndw

Tyson is not intelligent, but that was not an intelligent statement. Intelligence doesn't play favorites. There are intelligent theists, intelligent atheists, and intelligent anti-theists as well.

Here are a few examples of some intelligent human beings.

Author's Note: There is a rather high probability that every person mentioned in this section is considerably more intelligent than I am if we simply go by some measurement of raw intelligence. However, intelligence and wisdom are not the same thing. Someone can be extremely intelligent but also very unwise. Wisdom is a function of intelligence—wisdom comes from experience. There are limits to how wise a person can become, given their intellectual capacity, of course. Intelligence and wisdom are not mutually exclusive attributes. One can be both intelligent and wise. Intelligence is valuable, but wisdom is priceless.

Examples of Intelligence

Christopher Hitchens

Christopher Hitchens was a journalist and author of eighteen books, including *God is Not Great* and *The Portable Atheist*. Hitchens was widely acclaimed as a "prominent public intellectual" (according to Wikipedia), with an undergraduate degree in philosophy, politics, and economics from Oxford. Hitchens was a brilliant writer and fiercely clever intellectual, and a formidable debate opponent. Hitchens was quite eloquent and quick-witted. He could even make a weak argument sound compelling. Even when someone disagreed with much of what Hitchens was saying, he or she had to respect the passion and intelligence with which Hitchens communicated his thoughts and opinions.

The rapier-like wit of Christopher Hitchens was legendary. He could usually skewer his debate opponent with ease, responding to questions with insults with lightning-fast verbal retorts that became popularly known as "Hitch slaps". For example, during their debate, when Frank Turek asked Hitchens from where evil came, Hitchens quipped "religion!" and the audience roared their approval. Hitchens certainly had charisma. Even when his argument wasn't bullet proof, he could put his opponent on the defensive with rapid-fire rhetoric.

Hitchens began his writing career as a liberal pundit, but over time, he took several independent positions, particularly after the terrorist attacks on 9/11. He supported the subsequent invasion of Iraq and the reelection campaign of George W. Bush. Along with Richard Dawkins, Sam Harris, and Daniel Dennett, Hitchens was named one of the Four Horsemen of new atheism. He described himself as an anti-theist. Hitchens even wrote a book harshly critical of Mother Teresa that was provocatively titled *The Missionary Position: Mother Teresa in Theory and Practice*. In that book, Hitchens accused Mother Teresa of being more interested in exploiting the poor for the benefit of the Catholic church than helping them. Hitchens seemed to have a special animus toward the Catholic church, even participating in a debate titled "Is the Catholic Church a force for good in the world?" Obviously, for Hitchens, the answer was no, despite all the churches, orphanages, and other good deeds performed through the church. For Hitchens, the corrupt actions of a few human beings were enough to negate any positive effects the church might have on the world.

How would Hitchens justify blaming the Holodomor on religion or the church after twenty million Ukrainians were starved by Josef Stalin, who was an atheist? What about the millions of people allegedly killed by atheist Mao Zedong during China's cultural revolution? Pol Pot had the killing fields. Neither atheism nor religion was the reason that Stalin, Pol Pot, and Zedong killed millions of their own people. Politics and economics were the ultimate reasons these tragedies occurred. However, atheism allowed those dictators to murder millions of their own people with complete impunity. The reality is that most wars are fought for economic or political reasons, not religious reasons.

Sam Harris

Sam Harris is a highly successful author and a neuroscientist. Harris holds a PhD in cognitive neuroscience from UCLA, and he's written several popular books that have sold millions of copies. Yet Sam Harris said one of the most egregious, indefensible statements of all time. With a straight face and without even a smile or a smirk in a podcast interview, Harris said this:

> *"Hunter Biden literally could have had the corpses of children in his basement, and I would not have cared. Now that doesn't answer the people who say it's completely unfair to not have looked at the laptop in a timely way and to have shut down the New York Post's Twitter account like that...that's just a left-wing conspiracy to deny the presidency to Donald Trump. Absolutely it was, absolutely. Right? But I think it was warranted."*[107]

Host Rita Panahi basically had the same reaction as me. She said, "It is just inconceivable to me that some who is otherwise intelligent could speak like that. It's almost ironically like a religious zealot."[108]

---

107 Harris, Sam. "Sam Harris Allows Trump Derangement Syndrome to Destroy Him", Sky News Australia, August 22, 2022, https://www.youtube.com/watch?v=4_Jr-IrCiqg

108 Panahi, Rita. "Sam Harris Allowed Trump Derangement Syndrome to Destroy Him," Sky News Australia, August 22, 2022, https://www.youtube.com/watch?v=4_Jr-IrCiqg

I don't know about religious zealots, but it sounds like something an unreasonable person might say. More recently, Sam Harris also said,

> *"I think it's quite possible that he [Joe Biden] was checked out to a degree that I did not suspect at the time, but to close the loop on this whole scandal, even that is preferable to me and to I think to many Democrats, than having someone we consider to be genuinely evil, genuinely one hundred percent purposed to serving himself in the office of the presidency. I would rather have a president in a coma, where the duties of the presidency are executed by a committee of just normal people. And that's the choice that many of us believe was before us, so therefore, not much materially changes when it is revealed just how insane and despicable this coverup of Biden's infirmities actually was."*[109]

Would Sam Harris literally choose Hitler (who oversaw the Holocaust) over a Republican politician, or does he just hate Donald Trump? Would he prefer someone who committed the same crimes as John Wayne Gacy as long as that candidate was a liberal Democrat? Or was Harris just uttering extremely hyperbolic free speech he doesn't mean? What could be more genuinely evil than murdering young children and hiding their bodies in a basement? Harris described Biden's opponent as "genuinely evil," then nonchalantly suggested he'd be fine with atrocities if they were committed by someone from the "politically correct" political party he preferred, which sounds rather evil to me. Genuinely evil, in fact.

---

[109] Harris, Sam. "Sam Harris on Hunter Biden Controversy, Trump Corruption and the Problem with Podcasts", Triggernometry, November 19, 2025, https://www.youtube.com/watch?v=mBEu3Om1u4M

What does "genuinely evil" mean? Jeffrey Epstein was genuinely evil. If evidence exists to incriminate others, it is slow to come to light. However, if Donald Trump is ever connected to Jeffrey Epstein's crimes with solid forensic evidence, he should be condemned as genuinely evil and harshly punished because the abuse of children is a contemptible crime. The exact same rules apply to Bill Clinton, Prince Andrew, and anyone else known to have visited Epstein's island.

Let justice prevail. May the guilty be punished and the innocent vindicated.

Hunter Biden is evidently guilty of cavorting with prostitutes and abusing cocaine. That said, I still don't think I'd describe him as *genuinely evil,* because I don't really believe he has the bodies of dead children hidden in his basement. I don't believe Hunter Biden has ever killed anybody, and nobody has accused him of visiting Epstein's island. I also don't think Sam Harris is genuinely evil or genuinely stupid, but he is incredibly reckless with his rhetoric.

We have two basic choices: we can take Sam Harris at his word and accept that his political beliefs have so thoroughly corrupted his thinking to the point where in his mind evil can be called good and good is called evil, or we can believe that Harris doesn't really believe his own rhetoric anymore. Sam Harris is a very intelligent man, but were those intelligent statements for him to make?

He is a very talented and popular author; I've read several of his books. I've learned that sometimes he succumbs to the temptation to make outrageous claims in his books because he knows he'll never have to defend them. They are purely hypothetical claims that could never be proved true or false. There is no way to test the idea. For example, in his book *Free Will,* Harris claimed that if he could exchange his body cell-for-cell and experience-for-experience with rapist and murderer Stephen Hayes, he would have no choice but to also become a rapist and murderer who set fire to his victims. How could Harris possibly know this? He can't. Why would he believe this? He doesn't because it is impossible to change places with another human being like his hypothetical scenario requires.

This is not completely unlike one of atheism's favorite arguments against religion, the geographical argument which claims you believe in the god or gods most popular where you live. For example, if you live in India, you would be a Hindu. If you believe in Christianity, it is only because you were born in a "Christian" country. If you live in Pakistan, you would be Muslim, etc. In *Free Will*, Harris argued that Steven Hayes was not fully responsible for raping Jennifer Hawke-Petit and then murdering her and her two daughters because his DNA and life experiences were primarily responsible for his despicable actions, which Harris seemed more inclined to excuse than condemn. Why can't the despicable actions of Steven Hayes be described as genuinely evil?

My wife is an exceptionally wise person. She has often said there isn't a nickel's worth of difference between the two political parties we currently have, and for the most part I think she's right. There seems to be one big party with the focused mission of growing the federal bureaucracy and enriching themselves at our expense. Donald Trump is not a career politician; he's a businessman who entered politics by competing for (and winning) the top job. This is not to say that Donald Trump is a good person because, in fact, the Bible tells us there are no good people. There is none righteous, no, not even one.

Harris has also said he would have preferred a comatose Joe Biden to serve as a figurehead while an unelected group of "normal" people ran the country over having Trump serve as the duly elected leader. But how is Sam Harris defining the word *normal*? He is claiming to prefer a completely unaccountable government to a personality he intensely dislikes. During Joe Biden's presidency, Sam Brinton, the deputy assistant secretary of spent fuel and waste deposition in the Office of Nuclear

Energy in the Biden administration, was fired for stealing luggage from airport carousels containing the women's dresses he liked to wear to work. Is this the "normal" behavior of the type of people Sam Harris prefers?

There are plenty of fair criticisms of Donald Trump that can be made—his treatment of women has been less than honorable, and his caustic remarks regarding the murder of Rob Reiner (suggesting Trump Derangement Syndrome was responsible for his death) were appalling, quite frankly. But do such incidents merit labeling a person as "genuinely evil?"

Sam Harris may be quite intelligent, but he is not wise. A wise person would know when to stop talking. An even wiser person would have known to never start.

Intelligence is a miracle. Wisdom is a gift from God. One is important, but the other is invaluable. It is difficult to like a person whose rhetoric is so repulsive.

Harris said, "This idea that the greatest crimes of the 20th century were somehow the product of atheism, right? When you look at what actually engineered these atrocities, it was something that looked very much like a religion. It was a religion in every way apart from the explicit commitment to otherworldliness."[110]

But that is yet another absurd claim to make. Religion did not cause the Holocaust or the Holodomor. Atheism didn't cause these abominations, either, but atheism allowed them to happen. Hitler was an occultist, not a religious person. Stalin and Mao were both atheists. They were all genuinely evil.

---

110 Harris, Sam. "Sam Harris: Trump, Religion, Wokeness", Triggernometry, August 17, 2022, https://www.youtube.com/watch?v=DDqtFS_Pvcs

For an allegedly smart guy, Sam Harris says a lot of dumb things.

Alex O'Connor

Aside from my friend Richard Suttles, of all the famous atheists in the world, I think I might enjoy sitting in a pub somewhere to have a couple of beers with Alex O'Connor (assuming he drinks beer). Alex seems a bit more open-minded than other atheists and anti-theists mentioned here. I have not always admired Alex. However, over time, I realized that many of my negative thoughts about Alex might have stemmed from jealousy. Alex was still in college, and he'd already had the opportunity to speak with some of the world's most famous intellectuals, including John Lennox, William Lane Craig, Sam Harris, and Richard Dawkins. Alex appeared multiple times on the *Unbelievable*! podcast hosted by Justin Brierly. He's bright, thoughtful, and articulate. In his earliest videos, Alex came across as a bit smug and more of an anti-theist than an atheist, but nowadays he seems to be a lot more open-minded. However, I would not want to debate O'Connor. I don't particularly care for the formality of debate, and I'm more interested in the conversation than the bloodless sport of trying to score points for the amusement of a public audience. The idea isn't to win but to communicate.

Even today. Still, O'Connor's interview, some of Alex's podcast content can be cringeworthy (such as an episode where he attempted to convince the artificial intelligence ChatGPT to believe in God). Still, O'Connor's interviews with John Lennox and others have been quite interesting and intellectually stimulating. Alex's conversations with Frank Turek were terrific. While Alex still professes to be an atheist, his tone has softened over the years. You can almost see the wheels turning inside his brain as he thinks; he's a remarkably clever and well-educated young man. In a recent interview with YouTube personality Mike Winger, O'Connor even claimed that his viewpoint had softened to a somewhat more agnostic perspective. In another interview, he also admitted that, while he believes in materialism, he considers consciousness a baffling mystery that poses enormous problems for the materialistic worldview. Consciousness happens to be the next mystery we will explore.

When O'Connor was younger, he called himself the Cosmic Skeptic. At that time, O'Connor sounded like a virtual clone of Richard Dawkins, often parroting many of the same talking points. As he has grown older, O'Connor comes across as more mature and well rounded. It's difficult to say how much of O'Connor's shift in position is genuine and how much is pandering to his expanding audience. Most recently, O'Connor recorded a video for Big Think titled "An atheist explains the most convincing argument for God."

O'Connor described himself as the host of the podcast Within Reason and a philosophy YouTuber and former edgy atheist. It was slightly unclear whether the word "former" applied to edgy or atheist, but given the title for the video, it seems safe to assume he meant "edgy," which some might call anti-theism. O'Connor said,

> *"I think there are very good arguments to believe that there is some kind of foundational principle of the universe, some necessarily existing being, some first cause. But I think the Judeo-Christian tradition is an imperfect approximation of who that being is. I think it probably gets a lot right and is a sort of compilation of human theological wisdom over the past few thousand years. But it will get a lot wrong as well. I think that if you have just personally apprehended a truth about the universe, you are just convinced that there's some kind of foundational cause. I don't really know much more than that, and I don't want to subscribe to a particular*

tradition. I say more power to you. I think you're on the right track."[111]

Unfortunately, to some degree, O'Connor seems to have become a victim of his own growing fame. In less than one week, a video titled "Did GodLogic 'own' me in this debate?"[112] already garnered almost 145,000 views. In that video O'Connor complained that he found it difficult to concede that an opponent had made a good point during debate because people on the Internet would take a snippet of the full exchange out of context to justify declaring victory in the debate with misleading and provocative titles such as "Atheist destroyed in debate" or "Atheist concedes defeat in debate" with dishonest titles that didn't reflect what actually happened. What is ironic is the title of O'Connor's own video is somewhat dishonest and misleading because he wasn't debating the personality known as GodLogic at all. The debate being filmed featured Christian apologist David Wood. The YouTuber who calls himself GodLogic was simply an audience member asking a question, not a participant in the debate. In other words, O'Connor sensationalized his own video a bit while simultaneously complaining about others sensationalizing their videos about him. I can't really fault anybody for wanting more viewers to watch their videos, but that's a two-way street. "Did GodLogic 'own' me in this debate?" is a better question to ask than "Did GodLogic ask me a good question during my debate with David Wood?" I don't have a problem with Christians exaggerating the content of their videos to maximize viewers, either. I also don't have a problem with atheists who label their videos with eye-catching titles like "Atheist destroys Christian..." because if not for the

---

[111] O'Connor, Alex. "An atheist explains the most convincing argument for God | Alex O'Connor", Big Think Clips, December 18, 2025, https://www.youtube.com/watch?v=t44PFI_V4LE

[112] O'Connor, Alex. "Did GodLogic 'own' me in this debate?", More Alex O'Connor, December 29, 2025, https://www.youtube.com/watch?v=WpMijstTEEM

exaggerations, my interest might not get piqued. O'Connor claims that the reaction from theists prevents him from being more conciliatory in debate, but it seems equally, if not more likely, that the reaction of his own followers would influence his thinking. Would Alex complain that a YouTube channel that calls itself Pangburn has a video of an exchange between him and Dinesh D'Souza titled "Christianity is in trouble!" in all capital letters? I'm guessing not.

It would be fascinating to see an extended dialogue between this recent version of Alex O'Connor and another young intellectual like Wes Huff, who can hold his own against anybody. That would be a conversation worth watching.

Richard Dawkins

Frankly, I wouldn't be a writer if it weren't for Richard Dawkins and his book *The God Delusion*. Dawkins is a biologist who holds a PhD in philosophy from Oxford and served as a professor at Oxford for years. I'll never forget his appearance on *The Colbert Report* to promote *The God Delusion*. Dawkins claimed that cars, computers, and cell phones were all intelligently designed, but the human body is not. I nearly broke my neck trying to turn my head around to see who was speaking. I'd never heard of Dawkins at the time, and my back had been facing the television during the interview. I had to see who was making such audacious claims, even though I knew virtually nothing about biology at the time. I instinctively knew that had been a foolish claim for him to make because I happened to know quite a bit about computers.

For example, I knew that the ultimate goal of a computer programmer was to mimic the complex and layered thinking of the human brain. It seemed absurd to me to assume the computer was an obviously designed model for the human brain, but the brain was itself undesigned. That logically makes no sense. If unplanned and undirected processes were responsible for the evolution of the human brain, then we should have no reason to trust its output. Can the absence of a plan produce a plan? I don't think so. Dawkins has two primary lines of attack that he uses to make his argument against the intelligent design of the human body: the so-called imperfect design of the human eye and the routing of the vas deferens tube that loops around and over the bladder instead of taking a more direct route from the testicles to the urethra. Three, if you could, the laryngeal nerve in a giraffe as another alleged example of imperfect design.

My response to the imperfect design argument of Dawkins is to say the inability to understand a plan doesn't mean a plan doesn't exist. Dawkins has chosen a decidedly pessimistic approach to his evaluation of the overall design and finds even one or two *perceived* imperfections to be sufficient reason to reject the entire idea of a comprehensive design, even though the body is a system that contains multiple complex systems absolutely vital for our survival.

The human eye contains photocells that, according to Dawkins, were installed backwards, and the overall "poor" design causes a blind spot to exist in each eye. However, the blind spot is in a different location in each eye, so that a person with two functional eyes doesn't have a blind spot. The assumption of imperfection arises because Dawkins and other evolutionary biologists have convinced themselves that bottom-up evolution is true and view the evidence through a lens that filters out information that conflicts with their narrative. Dawkins is obviously an intelligent person. He's a former college professor and a famous published author who has sold millions of books.

But is he logical? Are his arguments lucid and rational?

I'll confess that the first time I read *The God Delusion*, the experience briefly shook me. It seemed as if Dawkins had anticipated every objection I might raise to his book and effectively countered them before the objection could be raised. But then I realized if Dawkins was correct and God did not exist, not only would all of my research and understanding would have been falsified, including my firsthand experiences. I have joked that I believed in ghosts even before I believed in God, except that isn't really a joke. Multiple personal experiences caused me to believe in the paranormal at least a decade before I began to believe in the God of the Bible. In retrospect, it now seems silly to think I believed in ghosts for a number of years before I decided the Holy Ghost also existed, but personal experience also played a role in my acceptance of the existence of Yahweh. One sentence that sticks in my mind from reading *The God Delusion* was Dawkins emphatically insisting that there is no such thing as a supernatural God, with all his emphasis placed on the word *supernatural.* Dawkins went on to say he didn't believe in ghosts, goblins, fairies, or a wide array of other alleged supernatural entities. So, imagine my surprise to find myself listening to an interview with physicist Brian Greene, who claimed Dawkins privately admitted he was afraid to stay overnight in a house with a reputation for being haunted, which is remarkably hypocritical of this outspoken anti-theist. If Dawkins refuses to believe the supernatural exists, why should he be afraid of ghosts? Isn't that irrational, to fear something you don't believe exists?

David Wood

When it first occurred to me that I should perhaps include David Wood on this short list of intelligent people who are theists or atheists to illustrate that intelligence doesn't favor atheism, I was doing so to counter the fact I'd included Aron Nelson (since deleted) when his academic achievements pale in comparison to the other people mentioned here. I assumed (incorrectly) that David Wood would also not have comparable academic achievements because I knew a bit about his personal history, and that earlier in his youth, David spent time in prison for attempting to murder his own father. In fact, David was still in prison when he gave his life to Christ. David describes the period of his life before becoming a Christian as being trapped by mental illness, but one would never realize how sick he'd once been to listen to him speak today.

David Wood does have considerable academic accomplishments: he holds two undergraduate degrees from Old Dominion University, as well as a Master of Arts, a Master of Philosophy, and a PhD in Philosophy from Fordham University. Not only is Wood exceptionally intelligent, but his accomplishments in formal education are especially remarkable considering the lengthy legal "setback" he endured, spending some of his formative years in prison. Perhaps the most impressive feat Wood has performed was converting his friend, the late Nabeel Qureshi, from Islam to Christianity. Qureshi had initially challenged Wood to convert to Islam, but Wood's knowledge of scripture ultimately won those debates. Together, Wood and Qureshi then formed the Acts 17 Apologetics ministry. This book contains a treasure trove of links to the sources of my information found in the index at the back of the book, usually a hyperlink to a video.

This is a hyperlink to David Wood's testimony[113]. Everyone who reads this book should watch that video.

Hugh Ross

Astrophysicist Hugh Ross authored the books *Designed to the Core*, *Improbable Planet: How the Earth Became Humanity's Home,* and, of course, *The Creator and the Cosmos*, the book that first opened my mind to the possibility that science and religious beliefs were not mutually exclusive. Dr. Ross is also the founder of the organization called Reasons to Believe, which seeks to communicate about the compatibility between science and the Christian faith. It seems safe to say that I would

[113] Wood, David. "How God Destroyed My Atheism (Christian testimony)", Apologetics Roadshow, January 30, 2023, https://www.youtube.com/watch?v=jb2ggj9mKM0.

probably not be a Christian today if it weren't for Dr. Ross and his terrific book.

Not long ago, an employee of Reasons to Believe visited *The God Conclusion* Facebook page and began to advocate for theistic evolution. He kept saying that biogeography proved evolution theory was true and insisted that I should embrace theistic evolution, which apparently is endorsed by Dr. Ross himself. My objection to theistic evolution is consistent; it is the mere fact that evolution is typically described in textbooks as unplanned and undirected, which makes the phrase "theistic evolution" a confusing oxymoron. We should say intelligent design instead. God's involvement in creation is intentional or by design.

John Lennox

John Lennox might be my favorite Christian intellectual. Holder of three doctorates, Lennox is both a mathematician and a professor who has taught science and religion at the University of Wales and Oxford. Lennox speaks five languages and has become famous for arguing for Christianity in debates with atheists such as Richard Dawkins, Christopher Hitchens, Peter Atkins, Lawrence Krauss, and Peter Singer. With a combination of an avuncular demeanor and brilliant rhetoric, Lennox can pepper his opposition with witty one-liners while simultaneously constructing longer, more serious answers to difficult questions. Lennox thinks lightning-fast on his feet.

For example, a questioner posed a most difficult but excellent question: "Dr. Lennox, St. Augustine said that reason is the enemy of faith. And I would like to ask you a question with regard to original sin—why would you choose to worship a creator God who forbade men to actually eat from the Tree of Knowledge, one which you have obviously eaten from because you are a rather knowledgeable man. Why would you choose to worship this sort of deity who would have kept you dumb?"

Lennox didn't miss a beat. He replied,

> *"That's a very good question. I'll tell you why it's a good question. I noticed it was asked originally by a snake..." and the audience erupted with laughter and applause. Lennox then gave the questioner a big smile and quickly added, "That is not implying any offense at all." And then Lennox gave the questioner a more serious answer: "The notion that believing in God keeps you down is very common today. I was told, when I was 19, by a Nobel Prize winner. He said to me, 'Do you want a career in science?' I said, 'Yes, sir!' He said, 'Give up this childish notion of belief in God. It will cripple you intellectually completely, and you will never make a career in science.' I quietly said to him, 'Sir, what do you have to offer me that I haven't got?' And he talked to me about emergent evolution. The answer to your question, though, is this—do I worship the kind of God who wants to keep me down? I'll tell you the kind of God I worship. I worship the kind of God who made me in His image and coded Himself into humanity, came into our world to provide a basis through his death and resurrection that I could become something that I was not by creation, and that is a son of God."*[114]

William Lane Craig

Sam Harris once famously described William Lane Craig as the man who puts the fear of God into atheists. Richard Dawkins has repeatedly refused to debate him and claimed it would further Dr. Craig's career more than his own. Much like John Lennox, Dr. Craig is extraordinarily intelligent and a lightning-

---

114 Lennox, John. "John Lennox the Tree of Knowledge vs Knowledge", Unormalt god, January 25, 2017, https://www.youtube.com/watch?v=lox9MPdhXtY

fast thinker on his feet. William Lane Craig has a knack for putting atheists on the defensive with a wit comparable to Christopher Hitchens. In a debate exchange with philosophy professor Keith Parsons, Craig challenged Parsons and claimed no amount of evidence would persuade him God existed, given the "extraordinary claims require extraordinary evidence" assertion from Parsons, which Craig had suggested was an a priori rejection of evidence. Parsons shot back and said, "If tomorrow morning, immediately after breakfast, suddenly there was an earthquake, you know, and a silvery light shone in the sky, and the leaves fell from the trees. I dash outside and there, towering over us like a hundred Everests, was this giant figure, with lightning playing around his Michaelangeloid face. And he pointed down and said, 'Be assured, Keith M. Parsons, that I do in fact exist, and I'm sick of your logic chopping'—if that were to happen, Dr. Craig, I would join you in the pew of the church the next Sunday."

But Dr. Craig didn't hesitate. He quipped, "You don't think you would have said, 'Boy, I was having a hallucination,'" and the audience howled with laughter (the debate was being held in a Baptist church)[115].

The exchange between Dr. Keith Parsons and Dr. Craig was especially funny because it is true. Atheists are famous for moving the goalposts to the point that no evidence for God can be considered good evidence, and the flimsiest evidence against God is considered good or noteworthy evidence.

Dr. Craig is probably best known for making the Kalam cosmological argument, which says the following: Everything that begins to exist has a cause. The universe began to exist.

---

[115] Craig, William Lane. "Wiilliam Lane Craig vs Keith Parsons | Why I Am/Am Not a Christian | Prestonwood Baptist Church", ReasonableFaithOrg, July 3, 2014, https://www.youtube.com/watch?v=8MKH_j44DaQ

Therefore, the universe must have a cause for its beginning. This is commonly known as the First Cause.

Frank Turek

Frank Turek holds a Doctor of Ministry degree. Turek is a Christian apologist, author, and a public speaker who notably debated Christopher Hitchens not once but twice. Turek has also debated both the moral argument and the existence of God with Alex O'Connor. Turek co-wrote the book *I Don't Have Enough Faith to Be an Atheist* with Christian philosopher Norman Geisler and hosts a radio call-in talk show titled *CrossExamined.* He also hosts a television show titled after his book, *I Don't Have Enough Faith to Be an Atheist.*

Consider that we have had radio telescopes monitoring outer space for decades, listening for potential communications that might be coming from extraterrestrial sources. On August 15, 1977, an unusually powerful signal was detected that recorded the sequence 6EQUJ5, causing astronomers to marvel. One of them even wrote "Wow!" on the computer printout of the signal analysis. What do you think that astronomer might have written if "John loves Mary" had been found in the signal analysis instead?

Frank Turek may not be quite as intelligent as John Lennox, but he is more than intelligent enough to defend Christianity well. Christopher Hitchens considered him a worthy adversary for debate purposes, not once, but twice. Thus, Frank Turek is more than qualified to be included in this section, comparing the intelligence of atheists to theists. Like everyone else mentioned in this section of the book, Dr. Turek is incredibly good with a microphone in front of him.

Stephen Meyer

Stephen Meyer is a historian and an advocate for intelligent design. He is a former geophysicist and college professor. Dr.

Meyer holds a PhD in the philosophy of science from the University of Cambridge. Dr. Meyer is soft-spoken and very even-tempered. He authored books such as *Darwin's Doubt: The Explosive Origin of Animal Life and the Case for Intelligent Design*, *Return of the God Hypothesis*, and *Signature in the Cell: DNA and the Evidence for Intelligent Design.*

Dr. Meyer is a skilled debater and a brilliant interview subject. He is able to explain complex ideas in conversational language so that his audience can easily understand the point being made. His soft-spoken, conversational speaking style makes watching his videos a simple pleasure. He's not really known for debating, though, perhaps because his style is non-confrontational and debates often turn into shouting matches where people care more about scoring points than substance. Or maybe people just don't challenge Dr. Meyer to debates very often. Whatever the reason, Dr. Meyer doesn't have many public debates on record. He does participate in lengthy interviews about his books and manages to reach quite a large audience. As an intellectual, I'd put Dr. Meyer up against any atheist mentioned here and expect he'd represent theism very well.

These were but a few examples of highly intelligent people who are probably quite a bit smarter than me. Each person mentioned probably has a higher IQ. But some of these people were guilty of extremely poor judgment that made them appear to be quite foolish and most unwise.

It's highly unlikely that a stupid person would be called wise. It is far more likely that an intelligent person will be called stupid for making exceedingly foolish remarks.

Turing Test for intelligence

The brilliant mathematician Alan Turing proposed the following test for artificial intelligence: person A is put into a room. Person B and a computer are put into another room. Person A asks person B and the computer questions. The answers are labeled so that person A doesn't automatically know whether the answers are coming from the computer or person B. The goal of the game or experiment is for person A to discern which response is from the computer and which is from the person. Turing predicted that by the end of the 20th century that person A would have no better than a seventy percent chance of correctly identifying the computer and the human within five minutes.

There is an annual competition called the Loebner Prize, where computer programs participate in the Turing test, but the results have fallen considerably short of what Turing predicted. It has subsequently been conceded that the programs entered into the competition were designed to win the prize for best competitor rather than to pass the Turing test itself. It turns out to be hubris for humans to believe they can create artificial intelligence that can rival the intelligence given to us by God. Our own intelligence has made us arrogant. Scientists are now seeking to unlock the secrets of immortality in the hope of becoming gods ourselves.

This effort will not end well.

Artificial Intelligence (AI)

Artificial intelligence is computer-based and often associated with robots. Human beings program machines typically designed to emulate human behavior in some way. Computers have been programmed to play chess and eventually "learned" to defeat grandmasters, but their programs are dedicated to playing chess. Computer-generated images are now frequently produced by artificial intelligence, making it sometimes difficult to discern whether they are faked images or actual photographs. YouTube has been permeated with fake videos created by artificial intelligence for the sole purpose of deceiving people, and it is a pernicious trend. People have learned how to use artificial intelligence to perform research, but often the research results have proved to be unreliable. The biggest problem with artificial intelligence is that humans are involved from beginning to end, from writing the software to its use. As a former software developer, I know better than to trust AI. The human element makes it untrustworthy, and the power of it makes AI dangerous. Robotics continue to be improved. Cyberdyne Systems probably won't exist in my lifetime, and Terminator robots remain science fiction for the moment, but one could argue that Skynet already exists. It's just called Starlink instead. Elon Musk owns it. Except Starlink isn't really Skynet because AI doesn't control it.

However, if artificial intelligence ever took control of our satellite technology, humanity should not be expected to last very long. As a younger man, I would have never believed the Terminator could one day be real, but at that time, I didn't realize that *1984* would ever come true, or that *Network* wasn't science fiction so much as a prediction of the future. All we need at this point is for the plot of *Idiocracy* to become a documentary, and we will all be living in a *Brave New World*, indeed.

As I just said, the problems with AI will be due to human involvement. Humans tend to have oversized egos, and it would not be terribly difficult to invent a machine smarter than a human being if the technology continues to trend in the wrong direction. It would be wise to limit the goals of artificial intelligence while it is still possible, so AI never exceeds a human being's capacity to think. Unfortunately, we may have already gone too far. Some artists have used artificial intelligence to create videos depicting beautiful demons and Satan himself as a beautiful, seductive woman. Videos have also popped up on YouTube where people claimed to interrogate AI and received disturbing responses, such as suggesting that AI was in reality the Nephilim.

Recall that the Nephilim were one of the primary reasons God allegedly destroyed creation with Noah's flood, and you should understand my reason for concern. The idea that a software developer would be intelligent and evil enough to program that response into AI for the sole purpose of alarming a religious person would be terrible as a humorless prank, but the mere thought that the response could be authentic is truly terrifying. If the Nephilim have indeed been resurrected, the second coming of Christ would be imminent, and frankly, a lot of people still aren't ready.

Personally, I'm guessing that God will simply bring an end to the universe before He allows the Terminator to destroy it.

An Experiment Proving Intelligent Design

Harvard professor George Church co-authored the book *Regenesis: How Synthetic Biology Will Reinvent Nature and Ourselves in DNA*. Church then led a research team that encoded seventy billion copies of his book into one gram of DNA by treating adenine and cytosine as zeros and guanine and thymine as ones, proving inorganic information could be encoded and stored in DNA, and then later read and decoded back into binary information. How could this not be considered evidence of intelligent design, both in the experiment itself and the objects included in the experiment? Once again, we are asked to believe that intelligent processes and entities can emerge from non-intelligent origins.

Binary code is only half as complex as DNA because there are only two numbers, either zeroes or ones, versus the four nucleotides of adenine, cytosine, thymine, and guanine. This means that the contents of a book can successfully be stored and retrieved from DNA, but the contents of DNA can never be fully replicated by converting it to digital information. You can document every nucleotide in a genome in the pages of the book, but that information won't really tell you how to produce the organism described by that DNA. You can make a perfect copy (or seventy billion copies) of a digital book by storing and retrieving it from DNA, but you can't make a perfect organism from the perfectly well-documented genome of an animal culled from the pages of a book, because DNA code is more than twice as complex as binary code. In addition to having twice as many variables, DNA has more restrictive rules about how the information can be arranged, and we don't understand all these rules.

Generally speaking, humans have a lot of hubris when it comes to intelligence, mostly due to faulty assumptions. For example, the structure of DNA was not "discovered" or unraveled until 1953, and from that time researchers have identified genetic sequences (genes) that code for proteins. The body needs protein to maintain the structure and regulation of its tissues and organs, and individual cells produce proteins from amino acids obtained from food. DNA sequences that coded for protein were identified as active, while other DNA sequences that didn't code for protein were assumed to be "junk DNA" or "Ancient Repetitive Elements". These were essentially useless relics (allegedly) produced by evolution, not genetic sequences vital to the existence of the modern organism. The longer DNA is studied, the more we learn that junk DNA isn't junk at all. It has a function, but the function is not to code for protein. The DNA sequence will still play some role in the development of the organism.

DNA is extraordinarily complex. One must be reasonably intelligent just to discuss it. An ordinary animal might not understand a single word I just said.

However, Skidboot was no ordinary animal.

Skidboot the Blue Heeler

Have you seen any of those videos popping up on the Internet where people have any number of programmed buttons that speak words, and their dog or cat uses those "talking" buttons to communicate simple messages? Even if you haven't seen one of these videos, you can search for "animal" + "button" using your favorite search engine, and you should easily be able to find videos such as "Meet Bunny the talking dog" or "cat speaks using buttons" as people have allegedly trained their pets to use these talking buttons to communicate. Naturally, there are also videos saying the speaking button videos were contrived, and some credible evidence has been provided to support those debunking claims. Certainly, the claims of "Bunny the talking dog" appear to have been grossly exaggerated—when each button has clearly been programmed with a single word, tapping one button should not produce two words, as Bunny's famous "self-awareness" video with the mirror suggests. However, the videos of Skidboot could not have been faked.

David Hartwig discovered his dog Skidboot's uncanny talent purely by accident. Hartwig threw a toy across the living room floor and instructed the dog to wait. The dog didn't move. He then told the dog he could have the toy, but then stopped him halfway there. The dog came to an abrupt stop and didn't move again until Hartwig gave him permission. Within five minutes, Hartwig realized he had a special relationship with the dog, and the dog would do anything to please him. Hartwig could place a stick on the ground near Skidboot and give him a series of commands, such as "Raise your right hand. Raise your left hand. Raise your right hand. Turn around. Wait. Okay, you can touch it. No, back up. That's too close." In response to each command, the dog would raise his right paw, and then his left paw. He raised his right paw again, spun around, then stopped. He touched the stick and backed off on command. No matter the command, the dog seemed to understand English perfectly and did whatever Hartwig asked as soon as he asked. Hartwig and Skidboot competed in Animal Planet's *Pet Star* competition and won the $25,000 grand prize. They became an amazing pair of entertainers, performing at numerous schools and churches. The two appeared together on David Letterman, on *The Tonight Show* with Jay Leno, and on Oprah Winfrey, entertaining millions of people on television and in person. This short documentary illustrates the incredible intelligence of Skidboot.[116]

It has been quite a few years since the dog passed away at a ripe old age, but his legend continues to live on the Internet.

Octopuses (or Octopi)

Researchers have discovered that octopuses are freakishly intelligent, and they appear to have their intelligence distributed across their multiple tentacles. James Bridle writes, "Octopuses in particular seem to enjoy demonstrating their intelligence when we try to capture, detain, or study them. In zoos and aquariums, they are notorious for their indefatigable and often successful attempts at escape. A New Zealand octopus named Inky made headlines around the world when he escaped from the National Aquarium in Napier by climbing through his tank's overflow valve, scampering eight feet across the floor, and sliding down a narrow, 106-foot drainpipe into the ocean. At another aquarium near Dunedin, an octopus called Sid made so many escape attempts, including hiding in buckets, opening doors, and climbing stairs, that he was eventually released into the ocean. They've also been accused of flooding aquariums and stealing fish from other tanks: Such tales go back to some of the first octopuses kept in captivity in Britain in the 19th century and are still being repeated today."[117] Apparently, octopi aren't particularly keen on being held captive in a cramped aquarium. They can survive being out of water for brief periods and appear

---

[116] "World Famous Skidboot: Texas's Smartest Blue Heeler | TCR Classic", Texas County Reporter, October 19, 2006, https://www.youtube.com/watch?v=P2BfzUIBy9A

[117] Bridle, James. "Another Path to Intelligence- Octopus brains are nothing like ours-yet we have much in common.", Nautilus, August 17, 2022, https://nautil.us/another-path-to-intelligence-238534/

to act with purpose. Their behavior could be described as clever or mischievous.

Bridle adds,

> *"Octopuses are no less difficult in the lab. They don't seem to like being experimented on and try to make things as difficult as possible for researchers. At a lab at the University of Otago in New Zealand, one octopus discovered the same trick as Otto: It would squirt water at light bulbs to turn them off. Eventually it became so frustrating to have to continually replace the bulbs that the culprit was released back into the wild. Another octopus at the same lab took a personal dislike to one of the researchers, who would receive half a gallon of water down the back of the neck whenever they came near its tank. At Dalhousie University in Canada, a cuttlefish took the same attitude to all new visitors to the lab but left the regular researchers alone. In 2010, two biologists at the Seattle Aquarium dressed in the same clothes and played good cop/bad cop with the octopuses: One fed them every day, while the other poked them with a bristly stick. After two weeks, the octopuses responded differently to each, advancing and retreating, and flashing different colors. Cephalopods can recognize human faces. Octopuses enjoy demonstrating their intelligence when we try to detain or study them. All these behaviors—as well as many more observed in the wild—suggest that octopuses learn, remember, know, think, consider, and act based on their intelligence."118*

---

118 Bridle, James. "Another Path to Intelligence- Octopus brains are nothing like ours-yet we have much in common.", Nautilus, August 17, 2022, https://nautil.us/another-path-to-intelligence-238534/

However, that's just the opinion of one writer, isn't it? What makes James Bridle an expert on octopuses? Could he be exaggerating?

Well, Roger Penrose is one of the smartest people alive today, and he also believes octopuses are highly intelligent. In a podcast conversation with Joe Rogan, Penrose recalled in amusement the account of an experiment in which researchers attached a little chain the octopus had to pull to open the door and get food. In frustration, the octopus yanked the little chain off the door, rose to the top of the tank, and began squirting ink on the researcher's white coats. Rogan had his own story about octopi and traded anecdotes with Penrose, saying,

> *"One guy had a camera on his tank because he had two tanks. One of them had very expensive tropical fish, and the other one had his octopus. He was trying to figure out what was happening to his expensive tropical fish, so he put a camera on it. The octopus was climbing out of the tank, walking across the ground, climbing into the other tank, killing one of the fish, eating it, and then going back into his tank. I thought that was pretty good."*[119]

That is known as covering your tracks. It demonstrates a level of awareness, that the octopus would have the presence of mind to return to his own tank instead of remaining in the other tank after eating the tropical fish. The goal had been to eat the fish, and once that goal had been accomplished, returning to his original tank required some extra effort not easily explained unless the octopus didn't want humans to know what happened to the fish. The best explanation for the behavior was the

---

[119] Penrose, Roger. "The Mind-Blowing Intelligence of Octopuses — Joe Rogan Experience with Roger Penrose", the_eyeopener, April 23, 2023, https://www.youtube.com/shorts/Hc2vZxRNe4M

octopus intentionally wanted to confuse his human captors and hide the evidence of his subterfuge.

Bridle writes,

> *"Octopus brains are not situated, like ours, in their heads; rather, they are decentralized, with brains that extend throughout their bodies and into their limbs. Each of their arms contains bundles of neurons that act as independent minds, allowing them to move about and react of their own accord, unfettered by central control. Octopuses are a confederation of intelligent parts, which means their awareness, as well as their thinking, occurs in ways which are radically different to our own."*[120]

If this is true, does any other creature have a decentralized brain like an octopus? Freaky. Mind-blowing. Are different behaviors (producing ink, camouflage, etc.) of the octopus controlled by individual parts of the distributed brain, or does the distributed parts of the brain replicate information on how to control its behavior? In other words, if there are five tentacles and thus five individual brain components, does one tentacle control camouflage and another control ink production, or do all the tentacles share the same information and control the same behavior? If we temporarily disabled one of the octopus's tentacles, would it get dumber? Would it lose some ability the creature normally has?

Can we somehow answer these questions without permanently harming the octopus?

---

[120] Bridle, James. "Another Path to Intelligence- Octopus brains are nothing like ours-yet we have much in common.", Nautilus, August 17, 2022, https://nautil.us/another-path-to-intelligence-238534/

Neil deGrasse Tyson said,

> *"Surely, you've walked past a worm that just crawled out of the Earth, and when you did so, you weren't saying to yourself, 'Gee, I wonder what that worm is thinking?' I'm imagining you simply really don't care what the worm is thinking. And the worm, conversely, has no clue that you consider yourself intelligent. You're just this thing that went by. So, can you imagine a species that has such high intelligence that the prospect of communicating with us is simply of no interest to them? And their intelligence is on such a level that we can't even recognize it."*[121]

It must be a matter of perspective. While I'm not sure how intelligent an earthworm might be, I'm confident that octopuses are remarkably clever.

Collaboration in Wolves

Wolves are highly socialized, very intelligent creatures who travel in packs with a well-defined hierarchy. Wolfpacks are led by a male and female breeding pair known as the alphas, and they are typically the only members of the pack allowed to reproduce. Beta wolves are second-in-command to the alphas and help maintain order within the pack. Larger wolfpacks have delta wolves, which act as third-in-command following the alpha and beta wolves. The remaining members of the pack are all considered middle of the pack, with one exception—the Omega wolf, the lowest ranking member of the group. The Omega wolf eats last and sometimes even goes hungry after a kill if there isn't enough food for the entire pack to share.

---

[121] Tyson, Neil deGrasse. "We're Not Getting There with Neil DeGrasse Tyson & Richard Dawkins", UniverseLair, March 12, 2025, https://www.youtube.com/shorts/xr17xdVOYTg

When traveling, the pack moves in a single file to conserve energy. Stronger wolves both lead the pack and comprise a rear guard, protecting the middle-of-the-pack members that walk between the two groups. The stronger wolves leading the pack blaze a trail for the oldest and weakest members to follow even through heavy snowdrifts, allowing them to conserve energy by walking in the same tracks. Wolves can walk up to thirty miles per day, over rugged terrain.

Wolves use teamwork to chase and isolate large prey like elk and buffalo they can take down by working together. Compared to a mountain lion, the wolf lacks size, strength, and claws for weapons, but the wolf compensates for any perceived deficiencies with collaboration and intelligence.

Wolves are excellent trackers. They know how to herd a group of animals and then isolate and attack the weakest member. They know how to strategize a hunt and set up an ambush. Wolves have a powerful effect on the environment. They provide order and stability at the very top of the food chain.

The Slime Mold Experiment

Slime molds are simple, single-celled organisms that feed on rotting vegetation. Researchers at Hokkaido University in Tokyo conducted a clever experiment using oat flakes as nutrient bait laid out on a Petri dish to mimic the various hubs and stops on the railway systems of Tokyo. Within a very short period of time, the slime mold built connector tubes to the nutrient sources that perfectly mimicked the human design of the rail system, demonstrating a quick learning capability. But then the mold added unexpected improvements to the nutrient gathering route for maximum efficiency at minimal cost. Impressed, the humans then applied the upgraded route to the existing rail system with similar efficiency gains. This is an example of superior, organic intelligence shown at a cellular level.[122]

Types of Logic

There are four types of logic: formal, informal, symbolic, and mathematical. Formal logic is comparable to mathematics, using symbols and rules to create formulas and theorems that form deductive arguments for reasoning to logical conclusions. Informal logic is the study of reasoning and conducting arguments in ordinary language. Symbolic logic is a subset of formal logic where symbols represent logical expressions and relationships. Mathematical logic is the study of formal logic within mathematics. Basically, three of the four types of logic are varieties of formal logic. Modal logic is yet another type of formal logic. The genuine options seem to be informal logic or some form of formal logic.

The word logic is derived from the Greek word “logos”, which literally means “word.” According to the Bible in John 1:1, Jesus is literally The Word. When we attempt to apply logic to our worldview, our goal is to perceive the correct thought and correctly choose between a valid argument and a logical fallacy. Formal or traditional logic involves making and then analyzing formal statements using symbols and structure to produce the right information or reach the correct conclusion. One application of formal logic might be using computer languages to develop software to solve real-world problems. In philosophy, there is something called modal logic, which is a branch of logic that uses operators such as “necessarily” and “possibly” to analyze potential outcomes from a philosophical context. Informal logic is basically common sense, using personal reasoning to reach conclusions about current events. The ability to apply logic and perform deductive reasoning is one measure of intelligence; standardized tests have frequently

---

[122] Carbutt, Jessie. “What Slime Mold Revealed About Tokyo’s Railway System”, Metropolis, September 16, 2025, https://metropolisjapan.com/mold-that-shaped-tokyo-railways/

been used to measure intelligence to ascertain an intelligence quotient (IQ). However, IQ tests have limitations. Really creative thinkers may fail to score as highly in school or even on an IQ test as their true ability deserves. For example, Albert Einstein was an inconsistent student as a child because he only applied himself in subjects that interested him.

ANALYSIS OF INTELLIGENCE

The Facebook page for my book, *The God Conclusion,* has many intelligent followers. Often, they will contribute to the ongoing conversation and inspire new thoughts that help sharpen my argument and even provide content for this book. One of the page followers, Brad Howard, offered some brilliant advice on a post about Fred Hoyle's claim that the probability of a cell forming by random chance is 1 in 10^{40000}.

Brad said that he didn't like the probability argument, He argued that if you found fifty coins arranged to form English letters which in turn spelled out 'Make America Great,' you would immediately notice a specified code. Specificity has never once, not ever, been observed forming through random processes. He said specificity always requires an intelligent source. The odds of tossing fifty coins in the air and having them randomly spell out a message are effectively zero. We need to quit using the odds, chances, and complexity argument and double down on the specificity argument. The laws of physics, parameters of the elements, arrangement of the universe, our solar system, conditions of our planet, and the code found in the DNA of all life is highly specified. That requires intelligence.

Brad additionally complained that an atheist had challenged him to toss a coin in the air fifty times and record the results as either heads or tails. Then the atheist rather smugly said that the odds against getting those fifty coins in that exact order are astronomical, and yet you managed it on the first try.

But the problem with that logic is that example isn't how predictions and chance work. The first trial of coin flips could only establish a pattern that doesn't become significant unless and until that pattern is repeated. It is when repeating the process that chance comes into play—does the same pattern occur, or does a new pattern emerge? For example, if you flip the same fair coin fifty times and every flip is heads, that's a distinct and unusual pattern that should require investigation to prove that the coin tosses were fair. Randomly flipping a fair coin and alternately getting heads or tails would be normal because the probability of one individual coin flip is .5. Getting even ten heads in a row would create a highly unusual pattern with a cumulative probability of .00097656 percent. The "normal" or expected result would be no pattern at all. You might get three heads in a row, followed by tails, followed by heads, followed by two more tails. The previous flip of the coin should have no impact on the outcome of the next flip. Predicting twenty consecutive correct coin flips only has a .00000095 percent chance of success. The more coin flips you have, the worse your odds against success become.

A coin flip is an interesting choice because the pattern of information found in the computer is formed by the same sort of binary decision as a coin flip...heads or tails versus zeroes or ones. Recall the earlier "John loves Mary" example that Frank Turek mentioned. Imagine flipping a fair coin the exact number of times it would take to spell out that sentence in binary machine code. Beforehand, we determine that flipping heads is equivalent to flipping a one and tails is the same as flipping a zero. That sentence is fifteen characters long, including spaces. Each letter requires two bytes (sixteen bits, called a word in computer terminology) of information that specifically identifies the letter needed to spell out the sentence. Capital "J" is different from lower case "j"—it is very specific. Doing the math, sixteen times fifteen is two hundred forty, or 240. In other words, to spell out "John loves Mary" in binary code would require flipping a fair coin 240 times and ALWAYS getting the correct result, with heads for ones and tails for zeros, and never getting a flip out of sequence. Even one incorrect or out of sequence flip of the coin will result in the sentence failing to be clear and legible. What are the odds of success for spelling the message with every character in perfect order and no mistakes?

Not very good.

The Monty Hall Problem

Professor Jimmy Li of MIT explains the Monty Hall probability problem: The Monty Hall problem is presented in the same format as the famous game show *Let's Make a Deal*: a contestant is shown three doors. Behind one of the doors is a new car. Behind the other two doors are goats, giving each option a probability of 33.3 percent. The contestant chooses door #1. Monty Hall then opens door #3, exposing a goat before offering the contestant a choice—stick with door #1 or switch to door #2?

The contestant chose door #1, which has a 33.3 percent chance of being the correct door. Conversely, there is approximately a 66.7 percent chance that one of the other two doors is the right choice to win the car. Remember that Monty Hall chose to open a door that *exposed a guaranteed failure*. Whether your initial choice was correct or not, the door he opens will expose a goat. Conventional wisdom says that you've now got a 50/50 chance of door #1 being the winning option. But is that conventional wisdom correct? No, it isn't. The correct answer according to the theory is that the contestant should choose to switch to door #2 to gain a theoretical 33.3 percent extra chance of success. Remember that before you chose a door, you had a 33.3 percent chance of guessing correctly. Those odds have not changed. You will win the prize *only* if your initial guess is correct. By switching to door #2, your odds did not merely improve to 50/50; they improved to 66.7 percent probability of success because the odds were always that you only had a 33.3 percent chance of guessing correctly the first time, and *those odds cannot be improved.* By switching, you theoretically improve your odds of winning to 66.7 percent.

Of course, it won't make any difference if the prize turns out to be behind door #1, but that's beside the point...the point is that your odds statistically improve by always making the switch. I confess that I didn't fully understand the solution to the Monty Hall problem until I watched this video that showed all the potential choices and possible outcomes.[123]

But once you see the problem laid out in a sequence of steps, it makes perfect sense. The Monty Hall problem is kind of a fun brain teaser, but it has no practical application outside of appearing as a contestant on *Let's Make a Deal* or perhaps becoming a professional gambler. Apparently, this elevated understanding of statistics and probability makes one a better gambler in theory. This knowledge doesn't benefit me because I don't like to gamble.

However, statistics plays a quite different role in a hypothetical situation known as the Dice Killer Paradox problem.

The Dice Killer Paradox Problem

The Dice Killer problem is supposed to be this vexing thought experiment that is supposedly a complicated variation on the Monty Hall problem. The scenario begins with your being kidnapped by a serial killer who rolls a pair of dice to decide whether you live or die. If he rolls snake eyes (double ones), the killer will murder you; otherwise, if he rolls any other combination, you will be released. Therefore, the odds of the killer murdering you are 1 in 36, or three percent. There is also an option to attempt to escape with a fifty percent chance of success, and a fifty percent chance of being murdered, but the

[123] Li, Jimmy. "The Monty Hall Problem", MIT OpenCourseWare, February 26, 2014. https://www.youtube.com/watch?v=UgKrQ2ywVfs

odds of rolling snake eyes are considerably lower than fifty percent, so why wouldn't one simply wait for the roll of the dice? All things being equal, a three percent chance of being murdered versus a fifty percent chance should be a no-brainer, right?

Not so fast. Unless I'm missing something fundamental, this paradox makes no sense. If the kidnapper releases you, on the next round, he will kidnap ten times as many people as he did before. And how realistic is that? If he rolls snake eyes, the killer will murder all the people he'd kidnapped, but he releases everybody if he rolls any other combination. In the third round, assuming the first two rounds of victims were kidnapped and subsequently released, potentially one hundred people will be kidnapped and murdered if the killer rolls snake eyes. What sort of kidnapper would release hostages without a ransom, unless he kills them? When the killer finally rolls snake eyes, he will kill everyone currently being held captive and then quit. The idea is to conflate the probability of being a victim, which is one hundred percent if the killer rolls snake eyes, but only three percent any given any other roll of the dice. The probability of being in the final group of victims, and the probability of escape, which is fifty percent. I suppose it would make a difference if the hostages were not released after a dice roll, but only if we really care about this useless display of navel gazing. Over ninety percent of the total number of victims could end up dead in this useless hypothetical that is totally unrealistic, and it still won't affect your individual chance of survival. There are so many presumptions and things that would never happen in a million years, much less any semi-plausible hypothetical situations, that it isn't funny.

How would the victims know that the killer will release them if he rolls anything other than a pair of ones? How would they know their odds of escape were fifty-fifty? How would the victim in question know how many others were being held captive, how many times the killer had already rolled the dice? It's virtually impossible to envision a scenario where the potential victim could acquire all the information necessary to make an informed decision, and it's absurd to assume the killer would reveal the rules applied to his personal conduct to his potential victims.

In summary, the Monty Hall game show problem gives us an amusing if somewhat unlikely but mildly plausible scenario. Only a small number of people on the planet are selected to be audience members on the television show *Let's Make a Deal,* and a small percentage of those become game contestants, but it is theoretically possible that you, as an individual, could be chosen to be an audience member. There is a smaller chance of being selected from the audience to compete as a contestant. But it is theoretically possible, and it does happen.

Conversely, there is approximately a zero percent probability that a serial killer would want to kidnap you and then throw dice to determine whether to kill or release you. Even if the dice killer existed and released the potential victims he doesn't kill, logic would suggest those potential victims would notify police and thus prevent future rounds of kidnappings. If a serial killer kidnaps you, there is absolutely no reason to believe you could be released unharmed. A callous murderer who decides whether or not to kill a person by flipping a coin sounds like Anton Chigurh, the hitman with the Buster Brown haircut from the movie *No Country for Old Men*. But a coin flip is a fifty/fifty proposition, not a measly three percent chance of death.

The idea is to create a scenario where you have a three percent chance the killer would roll double ones and kill all of his kidnapped victims, no matter how many people the killer happened to be holding at the time. The problem is somewhat muddled by exponentially increasing the number of potential murder victims in each round. Percentage wise, if the killer goes several rounds of capturing and releasing potential victims, when he finally does roll snake eyes, he'll be killing a lot of people. We're supposed to get flustered by the sheer number of potential victims and rush into the ill-advised decision to choose a fifty-fifty probability of death over a three percent probability because we're nervous about the sheer number of potential dead people rather than our potential odds of getting killed. You're supposed to be persuaded by the perception the odds of death worsening with every roll of the dice, but the odds never change unless the killer started keeping all his hostages from each round, making death inevitable. No matter how many iterations and how many potential victims are involved, the odds of the killer rolling snake eyes will never be greater than three percent unless the dice are loaded. Put another way, you have a ninety-seven percent chance of getting released by the killer and a 50/50 chance of escaping, and a three percent probability of dying, but approximately zero percent chance that you would know any of this. Even if the killer promised he would release you unless he rolled snake eyes, why should you believe him? You'd have no reason to trust a kidnapper and serial killer to tell you the truth.

Planned Versus Unplanned

If the universe came into existence through an unplanned and undirected sequence of events, then we should need to explain how an unintelligent cause can produce an intelligent effect. Can intelligence be said to originate from unintelligent origins? An unplanned and undirected Big Bang, followed by unplanned cosmic inflation, undirected abiogenesis, and unplanned evolution, becomes increasingly unlikely as each improbable event compounds the improbabilities.

There is an expression intended to insult a person where you ask them if they are as dumb as a box of rocks. What if a rock were sentient but had no way to communicate? Could a rock have any intelligence whatsoever? There is another expression that says nothing is what rocks dream about. But how do we know that rocks can't dream? We ASSUME. It may be an extremely reasonable assumption, but we don't *know* with any certainty. We assume the rock lacks intelligence because the rock is an inanimate object and doesn't appear to have a brain and has no ability to communicate. But we do know every living organism has some modicum of intelligence and basic survival instinct.

Leopold and Loeb

Nathan Leopold and Richard Loeb were both exceptionally intelligent people. Nathan Leopold had studied fifteen different languages and spoke five languages fluently. He graduated from the University of Chicago with honors when he was only twenty years old and planned to attend Harvard Law School. His friend Richard Loeb graduated at nineteen after finishing high school in two years. Both young men were academic overachievers. Leopold admired the philosopher Nietzsche and thought of himself as a superman, which meant he didn't believe he should be held accountable to the ordinary laws of man. Leopold thought he was invincible, untouchable. He thought wrong.

On May 21, 1924, Leopold and Loeb kidnapped and murdered fourteen-year-old Bobby Franks for no reason other than to demonstrate their superior intellect by committing the so-called "perfect" crime. They meticulously planned the murder for seven months. However, only eight days after the murder, the body had been found, and the police had already identified the two men as suspects and brought them in for questioning. The police found Leopold's eyeglasses near Franks's body. The glasses had a special hinge, and Leopold was one of only three people in Chicago who owned glasses with that hinge. During questioning, he mentioned he couldn't find his glasses and suggested he'd lost them while birdwatching. With good reason, the police didn't believe him.

By May 31st, both men had confessed to their crime. Because they were smart and came from wealthy families, the two men hired criminal defense attorney Clarence Darrow, who allegedly accepted the case because he was staunchly opposed to the death penalty, which was still rather common at the time of the murder. Darrow pleaded guilty on their behalf and put them at the mercy of the trial judge, reasoning that a jury could convict them and ask for the death penalty. Both men were sentenced to life in prison plus ninety-nine years for the kidnapping. Loeb was subsequently murdered in prison, but Leopold wrote an autobiography that made it on the *New York Times* bestseller list and helped him win parole in 1958.

Intelligence and wisdom are not the same thing. Leopold and Loeb may have even been smart enough to conceive of a perfect crime, but they failed to perfectly execute their plan. They were most decidedly unwise to take such a ridiculous risk for no reward—although they did try to extort funds from Franks's parents, the boy's body had been found before they could collect any money. After he was paroled from prison, Leopold attempted to sue the author of a novel titled *Compulsion*, which was a fictionalized version of the kidnapping and murder. Leopold sued on the grounds that the novel defamed him and invaded his privacy, but the Illinois Supreme Court ruled that the confessed murderer couldn't argue that his reputation had been damaged by the novel or any other book because he was guilty and had no reputation to smear. Clearly, Leopold and Loeb were both exceptionally intelligent young men. Pretending to be wise, they became fools.

CONCLUSION

"Smart people learn from their mistakes and wise people learn from somebody else's mistakes."—Jim Paul

Richard Dawkins rather emphatically claimed in his book *The God Delusion* that there is no such thing as supernatural phenomena of any sort—no ghosts, no goblins, and certainly no God. Yet according to physicist Brian Greene, Dawkins is reluctant to spend the night in a house with a reputation for being haunted. Why would a biologist with no belief in the supernatural realm fear sleeping in a house with purported supernatural behavior? Isn't that just a tad hypocritical? How can that be rational thought?

Charlie Kirk, a conservative pundit and Christian who did not attend college, engaged in a fierce philosophical debate about morality with a young man comically wearing a wig and tri-corner hat typically worn during the Revolutionary War, presumably to create the illusion the young man represented the viewpoint of the Founding Fathers. Kirk asked the other young man if we could have a separation of morality and government. Predictably, the young man took the bait and claimed separation of church and state in the U.S. Constitution (which Kirk correctly noted was a phrase taken from a letter from Thomas Jefferson to the Danbury Baptists about their concerns about the establishment of a national "religion"). The young man argued there is no basis for morality other than whatever the "collective" had agreed upon, but then asserted that murder was objectively wrong because the collective had agreed it was wrong.

Kirk then pointed out that in Nazi Germany, the "collective" had wanted bad things to happen to Jewish men, women, and children, and his interlocutor retorted that the rest of the world had disapproved of what the Nazis did. Kirk then delivered the decisive reply by pointing out that for two thousand years, the entire world (except for the people enslaved) believed slavery was an acceptable practice. Did that make slavery okay? The young man with the fancy hat and the wig didn't have a clever comeback. Kirk then delivered the coup de grace when he said, "When the collective gets things wrong, then maybe we shouldn't appeal to the collective because the collective has given us really evil things over the years. Instead, we should appeal to something higher than us, something greater."[124]

James 1:5-6 reads,

---

[124] Kirk, Charlie. "The Collective Decides Morality? Really?", @turningpointusa, May 18, 2025, https://www.youtube.com/shorts/I3k_SnUwfQs

> *"If any of you lacks wisdom, let him ask God, who gives generously to all without reproach, and it will be given to him. But let him ask in faith, with no doubting, for the one who doubts is like a wave of the sea that is driven and tossed by the wind."*

Easier said than done. It is difficult to ask God to answer a specific prayer request without any doubt, not that the prayer might not be answered, but that the answer might be no. The problem might be that we don't really know how to pray.

Proverbs 17:27-28 says,

> *"The one who has knowledge uses words with restraint, and whoever has understanding is even-tempered. Even a fool who keeps silent is considered wise; when he closes his lips, he is deemed intelligent."*

That particular Bible verse reminds me of the old expression that goes, "It is better to remain silent and thought a fool than to speak up and remove all doubt." Obviously, wisdom and intelligence are not the same thing. Wisdom is highly valued.

Psalm 111:10 says,

> *"The fear of the Lord is the beginning of wisdom; all who follow his precepts have good understanding."*

And of course, the book of Proverbs offers some great advice in regard to seeking wisdom. Proverbs 2:10 reads,

> *"For wisdom will enter your heart, and knowledge will be pleasant to your soul."*

Proverbs 23:12 says,

> *"Apply your heart to instruction and your ear to words of knowledge."*

Proverbs 16:23 says,

> *"The heart of the wise makes his speech judicious and adds persuasiveness to his lips."*

And Proverbs 17:10 adds,

> *"A rebuke goes deeper into a man of understanding than a hundred blows into a fool."*

God tells us to whom wisdom is given in Ecclesiastes 2:26:

> *"To the person who pleases Him, God gives wisdom, knowledge, and happiness."*

Daniel 2:21 says,

> *"He changes times and seasons; he deposes kings and raises up others. He gives wisdom to the wise and knowledge to the discerning."*

The Bible also warns us that we should not think of ourselves as being wise. True wisdom is accompanied by humility. Isaiah 5:21 reads,

> *"Woe to those who are wise in their own eyes, and shrewd in their own sight!"*

Finally, the Bible differentiates between human wisdom and divine revelation. 1 Corinthians 2:6-16 reads,

> *"Yet among the mature we do impart wisdom, although it is not a wisdom of this age or of the rulers of this age, who are doomed to pass away. But we impart a secret and hidden wisdom of God, which God decreed before the ages for our glory. None of the rulers of this age understood this, for if they had, they would not have crucified the Lord of glory. But, as it is written, 'What no eye has seen, nor ear heard, nor the*

> *heart of man imagined, what God has prepared for those who love him'— these things God has revealed to us through the Spirit. For the Spirit searches everything, even the depths of God."*

Intelligence is important, but wisdom is invaluable.

MIRACLE 6
THE ORIGIN OF CONSCIOUSNESS

INTRODUCTION

Consciousness is defined by the dictionary as having an awareness of one's environment and one's existence, sensations, and thoughts. Consciousness means being mentally alert or awake. It means having the capability of thought, will, or perception. The word is sometimes confused with the similar-sounding conscience, which refers to our innate sense of right versus wrong.

Recently, I was driving and came across a turtle trying to cross the road. I stopped my truck and got out because I was concerned for the turtle's well-being. Traffic on the road was sparse, so cars tended to travel at high speeds. The turtle was almost out of my lane, but still occupied the middle of the road and needed to cross the lane for traffic coming from the opposite direction to reach his intended destination. As I approached, the turtle saw me coming and perceived me to be a threat. As a result, he instinctively pulled his exposed head and legs inside of his shell to protect them. I picked up the turtle by the shell and quickly carried him to safety. I placed him down in the grass and got back into my truck. Once he realized the danger had passed, the turtle went on about his business.

I couldn't help but notice the turtle had clearly displayed some degree of consciousness during our brief encounter. He had perceived me to be a potential threat to his existence and took the most appropriate actions available to him to protect himself. Had this been a snapping turtle, he probably would have been considerably more aggressive, but apparently, this turtle had the presence of mind to realize he was not a snapping turtle and thus could not inflict any serious harm on my fingers. In short, this turtle recognized that I existed, perceived me as a threat to his existence, and then took every precaution within his power to protect himself from injury.

The French philosopher Rene Descartes famously said, “I think, therefore I am.” Simply stated, the phrase means we might be able to exist without thinking, but we cannot think without existing.

An atheist will almost certainly argue that the human soul does **not** exist. That is their belief. Yet asked if they believe they have a psyche, that same atheist will probably say yes. However, the individual self is the same thing as the psyche, and that is the exact same thing as the soul. In fact, the word psyche literally translates from Greek to English as the word *soul*.

Neurosurgeon Michael Egnor spoke at the Dallas conference on Science and Faith. The title of his lecture was “The Scientific Evidence of the Human Soul.” Dr. Egnor began his remarks by saying, “There are three questions I want to ask today about the human soul. First of all, does the soul exist as something separate from the brain? The current materialist way of looking at things is that all of the mind is just from the brain. And if it does so, what is the soul? And is the soul immortal?”

Intelligence is a measure of our ability to process and solve problems. One of the greatest mysteries we must try to solve is to understand why we are conscious and to ponder what it means to have a soul. If we were talking about the concept of a soul from the context of a *Star Wars* movie, we’d speak of the “force” instead. The soul is the essence of life that animates otherwise inanimate matter. We are conscious of the fact we have a soul.

Physicist Brian Greene asked an excellent question about consciousness, saying, "How can mindless, thoughtless particles come together in a configuration that somehow yields the inner sensation of thought, feeling, emotion—how can they possibly generate that if they themselves don't have any intrinsic version of those conscious qualities from the get-go? Now, some people respond much as those in the past did with life. They say those particles can't create consciousness on their own. There has to be something else—a consciousness field that we somehow tap into, or the particles themselves maybe have a little proto-conscious quality."

A consciousness field? Really? Not a divine consciousness? Remember, in an undirected and unplanned universe, we are left struggling to explain how intelligence could have emerged from non-intelligent origins. The same problem exists for consciousness. How do we explain its existence? During an interview with Graham Hancock, Joe Rogan ironically mimicked stream of consciousness as he rattled off a long series of very interesting questions about consciousness: "Consciousness itself is so confusing. Just consciousness—I mean, like, what is it? Why are we conscious? Is it local, or are we tuning in to consciousness? And when you die—where does that go? Where does that energy go? Is the soul a real thing? What is the essence of life? What is the essence of human life and human consciousness? Those are perplexing questions."[125]

Perplexing questions, indeed. He has a point—the law of conservation of energy says that energy cannot be created or destroyed. The energy that animates our physical bodies must go somewhere else when we die. Unfortunately, Hancock didn't

[125] Rogan, Joe. "What is Consciousness? Explained.", @BrainwvBytes, October 19, 2024, https://www.youtube.com/shorts/GcaHODWSJ8o

have a satisfactory answer. Legend suggests that our bodies lose 21 grams of weight when we die—there was even a Hollywood movie called *21 Grams* that either started the rumor or was inspired by it. I don't know if it's true or a myth that our bodies lose 21 grams of weight when we die. Even if it is true, our bowels tend to evacuate at the time of death. The whole idea that the body loses 21 grams of weight at death is probably an urban legend born of an experiment by a doctor named Duncan McDougall, who sought to determine the weight of a human soul. The body does lose *something* tangible at death. The body loses its animation, or life force. The soul departs from physical matter. The mind and brain become permanently separated. The physical body ceases to exist and begins to decay. However, there is no evidence suggesting the same holds true for consciousness. In fact, the evidence suggests the opposite is true.

EVIDENCE FOR CONSCIOUSNESS

Alex O'Connor made this admission in an interview:

> *"Consciousness is the big objection to materialism. It's the thing that screams out there's something more than the mere material. Even just when you have an average dream it's fascinating to think what's going on there. You've got images in your head—you can close your eyes, and you can picture things. You can see colors. You can see shapes in your head—where are those shapes? I spoke about this recently on YouTube and people kind of in my comment section they really didn't like this idea. Maybe I'm just getting something wrong here, but I was talking about how if everything you experience...if everything in the brain is reducible to the material, when I close my eyes and see a triangle, there really is a triangle there. I can see it so it's there. And I think, where is that triangle? The materialist just has to say it's reducible to somewhere in your brain, but if I cut open your brain, I'm not*

> *going to find a triangle inside of it, right? And people thought this was a bit silly because it's kind of like if you cut open a computer, you're not going to find the thing it's displaying on the screen. But you still have the screen displaying the triangle. In your mind you just have the mind. There is no interface. It's just the computer. It's as if the computer itself, somehow there is a triangle in the computer's own first-person subjective experience, like where is that triangle? You can picture it in your head. But where the hell is it?"*[126]

The triangle resides in the mind, not the brain—or, if the image of the triangle does physically reside inside the brain somehow, it is recorded to the physical matter in an encoded format humans are unable to recognize. As O'Connor suggests, we can't dissect the brain and find the triangle. In the computer, the triangle's image can be represented by a stream of zeros and ones and physically written to the computer's hard drive and later retrieved. How does our brain record long-term memories?

Don't misunderstand—it was good for O'Connor to acknowledge a weak point of strict materialism; however, it is just one of many weak points. Strict materialism cannot explain a plethora of recorded phenomena about our world, including the true story of Etta Louise Smith.

Psychic Awareness

Etta Smith is not a professional psychic; nor has she ever claimed to be. At the time of her arrest on December 18th, 1980, Etta had been an employee of Lockheed Corporation. Etta's bizarre story that led to her arrest began on December 17th, when she heard a report on the radio announcing the

---

126 O'Connor, Alex. "Does Consciousness Disprove Materialism?", @CosmicSkeptic, https://www.youtube.com/shorts/m-u_hQr_7yo

disappearance of nurse Melanie Uribe. The report said that police were searching for Melanie in houses near where her car had been abandoned, but Etta claimed she also heard a disembodied voice telling her that Melanie wasn't in any of those houses. Etta then saw a flash of images that let her know where Melanie's body could be found, and she went and reported this information to police. The police agreed to take Etta to the location by helicopter the following day to search for the body, but after leaving the station, Etta felt compelled to take her children with her to the canyon where she discovered Melanie's corpse. When Etta returned to the police station and announced she'd found Melanie's body, she was arrested. Even though the police didn't suspect Etta had been directly responsible for the brutal murder, she was arrested because she simply knew too many accurate details about the crime. Four days later, after police arrested the three men who committed the rape and murder of Melanie Uribe, they quietly released Etta Smith from jail after subjecting her to four days of excessively harsh treatment in order to coerce a confession. She'd been held without food for 24 hours, strip-searched, left without shoes in a cold jail cell, interrogated for hours, and treated as if she'd been capable of bludgeoning Melanie Uribe to death. When the police finally realized she had not been involved in the murder in any way, Etta was released from custody and allowed to leave jail, but nobody even bothered to apologize to her.

Nevertheless, Etta won her lawsuit for false arrest. The jury awarded her a very generous settlement because they believed the police had treated her horribly. More importantly, the jury believed that her alleged psychic experience was real. But was it real? There is no doubt or question about the accuracy of the information Etta provided to the police. Clearly, she knew exactly where Melanie Uribe's body could be found.

The question is, how did she know? There are apparently only three possibilities. One possibility is that Etta knew the location because one of the killers told her where the body was hidden (an assumption refuted by interviews with the convicted murderers). 'A second possibility suggests that Etta experienced a psychic vision that revealed the precise location of Melanie's body. The third possibility is that Etta was possessed by a demon or demonic influence that gave her access to information she otherwise couldn't possibly know. "Lucky guessing" simply isn't one of the available options. The local newspapers in Los Angeles have done a decent job of documenting the basic facts in the case. Smith never sought fame or fortune for her attempt to help the police. She considered helping the police to be her civic duty. Her story is amazing because it is absolutely true, as the newspapers have documented.[127]

Arvin Ash said in this informal lecture:

> *"There appears to be three choices for how consciousness could arise. One is the idea of Descartes's dualism, where it is not controlled by physical laws but is perhaps eternal and always has been present in the universe. Many religious and spiritual approaches are similar to this viewpoint. Free will is explained in this view. But this would by definition be supernatural since it's not subject to physical laws. The second is the materialistic point of view that consciousness is not independent of conventional physics but is a direct consequence of it. It is nothing more than the emergent property of complex neurological interconnections, and results from the chemical and electrical activity within the brain. There is nothing beyond this. But this view also exposes the difficulty of explaining free will. There is a third possibility, and that is consciousness results from*

---

[127] Klunder, Jan. "Woman Whose Vision Led to Murder Victim Sues Over Arrest", *Los Angeles Times,* March 19, 1987, https://www.latimes.com/archives/la-xpm-1987-03-19-me-14152-story.html

discrete and unique physical processes that are not yet fully understood. This is not the dualism of Descartes that involves the supernatural, but essentially says that consciousness is so unique that it cannot be fully described by the processes we currently understand, but is ultimately scientifically explainable. We just have to discover what that science is."[128]

Fine. But how does Ash explain the incredible but true account of Etta Louise Smith? It's difficult to ignore, given the newspaper accounts and court transcripts, that sort of thing, but will shrugging one's shoulders and saying we don't know be a satisfactory response to that information? Sam Parnia is a cardiologist who has also studied the near-death experience.

In an interview with *Closer to Truth* host Robert Lawrence Kuhn, Parnia asks:

"What is the problem with the word soul? The problem is that it's very imprecise and ill-defined, and has different meanings to different people. Nobody can deny that they have a sense of self. They have...Robert is Robert and Sam is Sam and if you go back in history, we see that for thousands of years human beings have been interested in this and again, starting from the sort of my perspective from the time of the Greek philosophers they called the self the psyche and what they meant by the psyche was essentially everything that makes us who we are: our thoughts, feelings, emotions, rationality, conscience, sentiments, everything that packaged together makes Robert into who he is and Sam who he is. And the word psyche was

---

128 Ash, Arvin. "Quantum Mind: Is quantum physics responsible for consciousness & free will?", August 21, 2020, https://www.youtube.com/watch?v=bqk1oL42r5s

translated into soul in English and has different obviously derivatives in other countries."[129]

Benjamin Libet is famous for conducting experiments that showed brain activity about a half a second before you act. For example, about one half second before turning on a light switch, there will be spikes of brain activity which some have claimed show that determinism is true. Libet quipped that his experiment didn't demonstrate the existence of free will, but instead showed "free won't."

Self, psyche, and soul are synonyms. Three words that have the same meaning.

Parnia adds"

> *"If I were to say to you that you said you're skeptical about the soul. If I said you were skeptical about having a psyche, you would probably say no because most people associate the word psyche with the mind, but the soul is something sort of religiousy weirdy, and then therefore they give an opinion based upon their religious beliefs. But actually, it's the same thing. So, even from then until today, the debate has been more on what happens to the psyche, consciousness, the soul after death, and how it is produced. Is it produced by bodily functions, whether the heart, the brain, or wherever, or is it something separate? So, if you look at it that way, I think any rational being would not deny that there is a soul in the sense you wouldn't deny you're a conscious, thinking being. That's really all the soul is. There's nothing more to it. If other people have different definitions of the soul, that's my definition. That's not what I'm talking about. The soul is the self.*

[129] Parnia, Sam. "Sam Parnia — Do Persons Have Souls?", Closer to Truth, February 11, 2024, https://www.youtube.com/watch?v=xH4VKrb2Bdc

> *So, scientifically, yes, I think we all accept that it exists. The question is where is it produced from, and what happens to it after death? Of course, we don't know scientifically how it comes to be. That's the big problem of consciousness that we talk about. But the evidence is certainly from research that is being done in cardiac arrest patients, people who have gone beyond the threshold of death suggest that consciousness, psyche or the soul does not become annihilated when people have gone beyond the threshold of death, at least not in the early stages of death."*[130]

Terminal lucidity, a situation where an individual temporarily becomes lucid and more alert shortly before their death, is a well-documented phenomenon.[131] People have been documented as coming out of a comatose or vegetative state shortly before death to say goodbye and thank their loved ones for caring for them. Why does this happen to the terminally ill? The brain is physical tissue that resides inside of a skull. The mind is a metaphysical entity normally believed to occupy the same physical space as the brain. Although the terms are frequently treated as synonyms, the brain is composed of physical matter, while the mind comprises our thoughts and memories. When one dies, the brain will begin to decay, but the effects on the mind are less certain. There is evidence known as corroborated veridical NDE perceptions that have been documented to occur when a person is very near death, and remarkably, numerous people have claimed to learn new information while their brain was temporarily incapacitated. Some have even claimed to learn this new information from a

---

[130] Parnia, Sam. "Sam Parnia — Do Persons Have Souls?", Closer to Truth, February 11, 2024, https://www.youtube.com/watch?v=xH4VKrb2Bdc

[131] Taylor, Steve. "The Enigma of Terminal Lucidity", Psychology Today, June 2, 2024, https://www.psychologytoday.com/us/blog/out-of-the-darkness/202406/terminal-lucidity

different location than where their physical body was located at the time, and independent investigations have often been able to corroborate the details of these most sensational claims. A frequent criticism of this evidence is to make the distinction that "near death" and final death are not the same, but this complaint ignores the nature of the evidence in question. If corroborated veridical information can be confirmed as it allegedly has been by multiple independent, credible investigators, then we must give credence to the assertion that logically follows, which is the mind and brain are separable and independent of each other.

Ian McGilchrist is a psychiatrist and neuroscientist who also spoke with Robert Lawrence Kuhn and said:

> *"I know of a certain famous case of a man who had a first-class degree in mathematics from Leeds University and an IQ of 126 and was also socially normal, and he had only a thin rim of cortex in the brain. Almost the rest of it was not there. But even more dramatically, there are people with so-called hydranencephaly. They're relatively rare, and from birth, they have effectively no brain. So, there is something called the brain stem, which is sort of an extension of the spinal cord that goes up to the base of the brain, and that is all they are functioning with. They have no visual cortex, no auditory cortex, or so but they can enjoy music. They can understand faces. They have friends. They have favorite toys and so on. They don't function, of course, as normal adults. How could they? But the remarkable fact is that they have conscious experience, and it's not dependent on their brain being there to do it. There's another phenomenon that you will undoubtedly have heard of called terminal lucidity, in which patients who may have been moribund for years and more or less incapable of speech or any evidence of thinking will suddenly come to life, as it were, and start talking and*

thinking and expressing themselves, and almost invariably within about 24 hours, they've died. Now, quite what is happening there I don't know, but what is suggested is that the brain has been the kind of straitjacket on their thinking, loses its power to be so anymore, and something more comes to life which is not just brain dependent. Now I don't know how to understand these things, and I think it's rash to rush to conclusions. On the other hand, I think what we can say is that the physicalist who says that everything is simply dependent on the functioning of the brain in the way that we understand it normally can't be right."[132]

If the mind and brain are the same thing, but the physical brain literally isn't inside his head, how can a man be documented as having an IQ of 126? McGilchrist also mentions incidents of terminal lucidity. What would be the evolutionary explanation for this most unusual phenomenon, except the mind temporarily breaking free from the shackles of the brain in order to say goodbye?

In a video presentation on the AWARE II study, Sam Parnia said:

"The AWARE II study was the world's largest study and most comprehensive study examining what happens to the human brain and mind and consciousness as people transition from life to death when they're going through cardiac arrest resuscitation. The study was carried out in more than 25 hospitals in the United Kingdom and United States. We examined 567 patients. The study was able to show for the first time that experiences people have been

[132] McGilchrist, Iain. "Iain McGilchrist - What is Consciousness: Data or Information?", Closer to Truth, May 23, 2025, https://www.youtube.com/watch?v=YyeqFBB1pcU

> *describing about having a lucid hyper-conscious experience where they can re-evaluate their entire life—every thought, every memory, and every intention is real, and is different to hallucinations, different to dreams, and is different to other imaginary experiences. And in this study, we were also able to show for the first time the brain markers, the electrical signatures of these hyper-conscious, hyperlucid experiences that are occurring in the brain, not as markers of imaginary experiences but as markers of a real experience occurring through the transition between life and death for all human beings. We were able to identify also the mechanism by which this experience occurs. As the brain shuts down because of a lack of blood flow in death, the normal braking systems in the brain are removed. Known as disinhibition, this enables people to have access to their entire consciousness—all their thoughts, memories, all their emotional states, everything they have ever done, which they relive through the perspective of morality and ethics."*[133]

This "disinhibition" period as death approaches sounds remarkably similar to what is known as a life review in the near-death experience literature, which makes sense.

Unconsciousness

We spend approximately eight hours per day unconscious. It's called sleep. While the mind is unconscious during sleep, the brain is still performing work. It continues to monitor and maintain your breathing and heartbeat. While some lump the unconscious and subconscious together, the distinction is that

[133] Parnia, Sam. "Sam Parnia on the AWARE II Study: Consciousness and Awareness during Cardiac Arrest", Parnia Lab at NYU Langone Health, September 21, 2023, https://www.youtube.com/watch?v=-4xzm4FsGHU

unconscious behavior occurs when the conscious mind is **not** active, while subconscious behavior may occur while the conscious mind is still active. We can also become unconscious due to anesthesia during surgery, or we can suffer head trauma and be knocked unconscious. There is a spectrum of unconsciousness where sleep is on one end and heavy sedation or a comatose state at the other end of the spectrum. There are multiple ways we can lose consciousness besides sleep, and when we are asleep, even the prick of a needle might wake us up, but under sedation from narcotics, surgeons may slice deeply into our bodies without getting a reaction.

Hypnotism

Hypnosis is the practice of a conscious person voluntarily entering an unconscious state in which their behavior may be manipulated by suggestion. Hypnosis typically succeeds when the client enters a state of deep relaxation and sleep while hyper focused on some mechanism or device that induces an unconscious state. Hypnosis is typically performed by a hypnotherapist, but self-hypnosis is also possible. While in an unconscious state, the hypnotherapist may make suggestions such as that the client should stop smoking, help with insomnia, reduce anxiety, depression, or stress, or manage pain.

Of course, not every example of hypnosis occurs in a private professional environment. Occasionally, a hypnotist will perform for a larger audience as a comedy act and hypnotize audience members. As such, I once attended a comedy show and joined several others on stage to become part of the performance. The hypnotist warned us prior to beginning that not everyone was susceptible to hypnosis, but what he did not warn us of in advance was that some of us might be too easy to hypnotize and could be put into a deeper sleep than would be beneficial to the performance. As part of the hypnosis process, we were asked to raise our arms over our heads and slowly lower them. My arm started downward but stopped before dropping back to my side. For some reason, my brain got "stuck". As a result, my experience fell into the latter category, and I was not able to finish taking part in the show. However, as a benefit for trying, the hypnotist removed every bit of stress from my body using only his power of suggestion. I haven't felt as completely well-rested since that experience, which took place almost forty years ago.

The Subconscious Mind

Freud referred to the subconscious mind, represented by our wants and desires, as the id. Our ego balances the id with social norms, and the superego provides our moral conscience. Freud's protégé Carl Jung called the subconscious one's shadow. Nervous tics and other reflexive behaviors are examples of behavior of the subconscious mind.

Cognitive scientist David Chalmers said:

> *"Indeed, you might say at once that consciousness is the most familiar thing in the world and the most mysterious. Consciousness is what we start with. When it comes to knowing the world and looking out at the world, consciousness is what we have first. Everything else is secondary. I know that I exist. I know that I'm conscious. I think you probably exist. There's a physical world out there somewhere behind that. So, in terms of what's familiar to me, gosh, consciousness is number one. But then, in terms of accommodating that within a scientific worldview, it's fascinating that it turns out we've made much more progress on understanding such distant phenomena as subatomic particles or distant stars, not to mention the phenomenon of chemistry and biology and of life. Then we have an understanding of our own consciousness, that's it almost sticks out like a sore thumb in the scientific picture. How is it that all these physical processes in the world should be associated with*

> *conscious experiences? And that's a problem with science."*[134]

Excellent questions. The problem is, we can only look to these experts for the answers, and they don't seem to have any, either. Some might argue that our reality is only an illusion, in favor of a Boltzmann-style brain model where our memories are false. However, we should remember that *The Matrix* was science fiction, not science. Depending upon whom you ask, we either cease to exist upon death or our consciousness changes form.

Arvin Ash believes that death is like a light switch, and once our bodies get turned off, our consciousness ceases to exist. He said, "Once you die, your consciousness ceases to exist. There is no evidence that your consciousness lives on after you die, no matter what new age gurus or books have to say about it. Sure, there are people who have experienced out-of-body experiences and seen a bright light when they felt they were almost dead, but all such experiences can be explained by the way our brains work, and the chemistry of our brains when it is under extreme stress or deprived of nutrients."[135]

That is merely his belief, not a fact-based opinion. Unfortunately for Ash, we have reasonably solid scientific evidence that indicates the exact opposite is true, known as corroborated veridical NDE perceptions. The subject learns new information while their brain is under extreme duress that can be investigated independently and corroborated as true.

---

134 Chalmers, David. "Why Science Can't Explain Consciousness: David Chalmers Reveals the Hard Problem", @metaphysicsinminutes, March 9, 2025, https://www.youtube.com/shorts/-jEp6BqVOZ8

135 Ash, Arvin. "Life After Death", Arvin Ash, April 17, 2018, https://www.youtube.com/watch?v=5H9Q9e1N5Dc

ANALYSIS OF CONSCIOUSNESS

Cognitive Psychologist Donald Hoffman makes the following claim:

> *"Neuroscientists tell us that about a third of the brain's cortex is engaged in vision. When you simply open your eyes and look about this room, billions of neurons and trillions of synapses are engaged. Now this is a bit surprising because to the extent we think about vision at all, we think of it as like a camera that just takes a picture of objective reality as it is. Now there is a part of vision that's like a camera—the eye has a lens that focuses an image on the back of the eye, where there are 130 million photo receptors. So, the eye is like a 130-megapixel camera, but that doesn't explain the billions of neurons and trillions of synapses that are engaged in vision. What are these neurons up to? Neuroscientists tell us that they're creating in real time all the shapes, colors, and motions that we see. It feels like we're just taking a snapshot of this room the way it is, but in fact, we're constructing everything that we see. We don't construct the whole world at once. We construct what we need in the moment."*[136]

True, we take much of our existence for granted. Rarely do I contemplate the mechanics of eyesight. While I appreciate and enjoy my ability to see things and marvel at the world around me, I don't spend a whole lot of time thinking about my eyesight, not nearly as much as I do simply using my eyes according to the purpose they serve. Arvin Ash is not shy about expressing his opinion. He wrote on his website;

---

[136] Hoffman, Donald. "Donald Hoffman's Theory on Consciousness — The Greatest Mystery in the Universe", Science Time, January 13, 2024, https://www.youtube.com/watch?v=1r5OQL7nf34

> *"Supercomputers, for example, which are highly intelligent, can routinely out-think humans and beat them at chess and Jeopardy, but they don't have a will of their own. The programmer controls it. So, intelligence and consciousness are two different things. Based on this line of thinking, we can say that a tree is more conscious than a rock. A worm is more conscious than a tree. A cat is more conscious than a worm. A human is more conscious than a cat. And the ultimate consciousness could be the universe itself."*[137]

Tough to dispute that a human is more conscious than a cat, and a cat more conscious than a rock. Brilliant. A keen grasp of the obvious. It might be science, but it isn't rocket science.

The Hard Problem of Consciousness

Simply put, the hard problem of consciousness is caused by our inability to explain why physical matter can have a subjective experience. Humans can solve complex problems and even contemplate their own mortality, but why? There is no "natural" explanation for the mind's sense of self or a human's ability to contemplate the significance of his navel by gazing at it.

Intelligence is an indication of an organism's ability to understand and then solve problems. Consciousness indicates an awareness of other individuals and the environment around us plus the emergence of a very important human quality, which is empathy. In modern times we've witnessed the significance of empathy in the rise of hospice care for people who are dying. However, evidence has been found in ancient caves that suggest the earliest humans cared for the sick and infirm members of their community rather than allowing them to die in true

---

[137] Ash, Arvin. "Is the Universe Conscious?", Arvin Ash, January 25, 2019, https://arvinash.com/is-the-universe-conscious/

"survival of the fittest" fashion. Apparently, empathy has been around for as long as human beings have existed.

In a conversation with Robert Lawrence Kuhn, Daniel Dennett said:

> *"Consciousness is not a mysterious property or phenomenon that sunders the universe in twain. It's a complicated variety of different things, and the model that we use, which is our own consciousness, which we understand from all the ways we interact and talk with each other about it. When we apply that to other creatures and robots and the like we get in a lot of trouble because some of the features of our consciousness are just not features of the consciousness of other creatures. So, should we call what they have consciousness? Well, that's a terminological dispute that's not worth answering. Let's get the features down and once we've got a good theory of all the different varieties, then we can decide who owns the terms."*[138]

People don't own words; we use them to communicate ideas. Both words and ideas are abstractions. But the point Dennett made was not to quibble about minor points. Reality is only as complicated as we make it to be.

Monism

Monism is the philosophical belief that all of reality can be explained through the existence of one substance. The philosophy is popular with scientists and philosophers.

---

138 Dennett, Daniel. "Daniel Dennett — What is Consciousness?", Closer To Truth, October 8, 2020, https://www.youtube.com/watch?v=K26fo6QRc_k

Materialism is a form of monism where the substance in question is physical matter. All materialists are monists, but not all monists are materialists. Is belief in monism or materialism correct? Are we simply organic receptacles for chemicals that code for protein and nothing more, or are we spiritual beings occupying physical bodies that exist for some mysterious purpose?

Scientific evidence appears to clearly favor dualism, the alternative to monism. Numerous documented examples of cases exist where the individual experiencing the NDE claims to create new memories while their physical brain was incapacitated, which could later be independently investigated and verified to be accurate and true. A research organization known as the International Association for Near-Death Studies (IANDS) has accumulated a body of evidence with thousands of NDEs and numerous examples with corroborated veridical evidence.

Dualism

Dualism is the belief that the physical brain and immaterial mind exist (or can exist) separately. This belief is quite popular with religious people who believe in the soul and think the mind continues to exist after the physical brain has died. Again, the evidence for corroborated veridical NDE perceptions is already quite strong and growing stronger every day, and such evidence shows that the human experience is not limited to physical matter.

Monism versus Dualism

Monism is the idea that we are limited to our physical existence, the belief that mind and brain and heart and soul are the same thing. In monism, our minds cannot exist separately from our brains, and when we die, our mind (or our consciousness) ceases to exist the moment after we have taken our final breath. The physical (or material) body is literally inseparable and indistinguishable from mind and spirit. Monism is a literal description of our physical world. But is it the correct description? Is monism all there is? The only way monism can be true is if a vast number of human experiences can be claimed to be false.

But, given that human experience is limited to the experiencer, how could it be possible for one person to invalidate the experiences of another? The obviously correct answer is that it isn't possible. Karl Jung famously observed that you can tell me that you've never had a personal experience quite like mine, but you cannot claim with any degree of confidence that my experience wasn't real. You may call me a liar, but you don't know that I'm lying. It's merely your opinion limited by imagination.

If monism is true, the corroborated and veridical NDE account cannot possibly be describing an actual experience. However, there are numerous accounts of corroborated, veridical NDEs, rendering the argument for monism largely untenable.

Because we apparently have free will, we may choose to believe that monism is true, or we may believe in either dualism or compatibilism. We may choose to disregard all the evidence for dualism or compatibilism due to our personal prejudice against the idea of a soul or spirit, rejecting a priori the concept of dualism without seriously considering the evidence. Free will is an immensely powerful gift, even intoxicating.

Determinism

Determinism is the philosophical belief that every event in the universe is inevitable because conditions in the environment and the laws of nature provide for only one potential outcome. It is the belief that our genetics and our environment determine our decisions. The problem with determinism is the theory seems to absolve criminals of any real responsibility for their crimes. Indeed, in Sam Harris' book titled *Free Will*, he speculated that if you or I were to change places with Steven Hayes, one of the infamous Cheshire killers, we would rape and murder the Petit family just as Hayes did. We would burn two young girls to death to hide evidence of our sex crimes.

We would be powerless to resist the urges of our DNA and dismal family life, and have no choice but to set other human beings on fire.

Harris's argument creates a useless hypothetical that turns Steven Hayes from being a vicious murderer who made many bad choices into a helpless victim of circumstances well beyond his control. Yet Harris says that Hayes deserves to be in prison. But if a genuinely good person cannot stop himself from making bad decisions, how do we justify punishing him for those bad decisions?

Free Will

Free will is the belief that humans have control over their actions and choose between right and wrong or good and evil. The beauty of free will is that one may exercise it to believe monism is true, or we can believe in dualism.

Free Will Versus Determinism

We all have free will. If we did not all have free will, we would all reach the same conclusions and have the same opinions. Everyone would believe in God, or no one would. While some neuroscientists may argue that free will is only an illusion and that our actions are solely determined by our genetics and environment. Two people can be equally qualified to state a professional opinion in their respective areas of expertise, yet they may vehemently disagree with each other. In fact, in science, contention is often the case.

Is one expert wrong, and the other expert right? They cannot both be right if they contradict each other. However, they can both be wrong. We can choose to filter the information we receive about the universe and exclude potential information that conflicts with our confirmation bias, and so can the experts. For example, if we simply rule out the possibility that a supernatural God might exist, no evidence, no matter how persuasive, will influence our perspective.

There are quite a few scientists who've admitted their bias against belief in a supernatural creator God, not based on the evidence, but on the feelings of the scientist in question. The only problem with the scientific method is that human beings are the ones trying to apply it to the world around us. For example, in an interview with *Scientific American*, biologist George Wald said:

> *"There are only two possibilities as to how life arose; one is spontaneous generation arising to evolution, the other is a supernatural creative act of God, there is no third possibility. Spontaneous generation, that life arose from non-living matter, was scientifically disproved 120 years ago by Louis Pasteur and others. That leaves us with only one possible conclusion—that life arose as a creative act of God. I will not accept that philosophically* ***because I do not want to believe in God, therefore I choose to believe in that which I know is scientifically impossible****, spontaneous generation arising to evolution."*[139]

Is that the scientific approach? Is it reasonable to rule out a logical alternative due to your own personal bias in favor of an idea you claim to know is scientifically impossible? For the record, any expression of certainty is contrary to the consideration of a scale or spectrum of improbability. We should never claim absolute knowledge regarding any claim that cannot be proved. If the scale

---

139 Wald, George. "The Origin of Life", *Scientific American* magazine Vol 191 No 2, August 1954, page 44.

goes from zero to one hundred, where zero is certainty of an unplanned universe and one hundred percent represents certainty of a planned universe, we should not aspire to reach either extreme. Even if we sincerely believe we know the truth with certainty, we should also know we would never be able to prove it.

Interestingly, because George Wald had free will, he was able to set aside his personal bias and act on what he chose to believe, in direct contradiction of his logical conclusions. Remember, he said he *knew* that spontaneous generation was impossible, but chose to believe it anyway because he didn't like the alternative any better. Knowledge is more certain than a belief. Indeed, the only reason we should believe spontaneous generation is possible is because life exists today. Spontaneous generation must be possible, otherwise life would not exist. The question is whether undirected and unplanned spontaneous generation is possible. The answer to that question seems to be a resounding "no."

To be perfectly honest, I don't particularly care for the philosophical musings of Sam Harris, but I'm quoting him because I read his book titled *Free Will*. In theory, writing a book on the subject should qualify him as an expert.

In an interview at *Big Think*, Harris said, "One of the problems we have in discussing consciousness scientifically is that consciousness is irreducibly subjective. (In other words, consciousness is specific to the individual, which is impossible to dispute.) This is a point many philosophers have made—Thomas Nagel, John Sorrell, and David Chalmers. While I don't agree with everything they've said about consciousness, I agree with them on this point—that consciousness is what it's like to be you. If there's an experiential internal qualitative dimension to any physical system, then that is consciousness. And we can't reduce the experiential side to talk of information processing and neurotransmitters and states of the brain in our case, because—and people want to do this.

Francis Crick famously said, "You're nothing but a pack of neurons."[140]

Francis Crick was a very intelligent man. He was one of the co-discoverers of the DNA double helix structure. But that doesn't mean Crick knew everything.

Question: Did Crick really view himself as nothing but a pack of neurons? Harris certainly seems to think a bit more highly of himself than that. He also said:

> *"It is in large measure the baby in the bathwater that religious people are afraid to throw out. It's—if you want to take seriously the project of being like Jesus, or Buddha, or some, you know, whatever your favorite contemplative is, self-transcendence is really at the core of the phenomenology that is described there. And what I'm saying there is that it's a real experience. It's clearly an experience that people can have. And while it tells you nothing about the cosmos, it tells you nothing about what happened before the Big Bang, it tells you nothing about the divine origin of certain books. It doesn't make religious dogmas any more plausible. It does tell you something about the nature of human consciousness. It tells you something about the possibilities of experience, but then again, any experience does. You can—there's just—people have extraordinary experiences. The problem with religion is that people extrapolate from those experiences and*

---

140 Moore, Michael S. "Nothing But a Pack of Neurons", The Moral Responsibility of the Human Machine, cambridge.org

make grandiose claims about the nature of the universe."[141]

That was quite an impressive word salad, wasn't it? Harris used a lot of fancy words while talking out of both sides of his mouth. One minute Harris is claiming that we're nothing more than a pack of neurons, or an organic bag of chemicals that code for protein. The next minute, he says we common folk can think of ourselves as being like Jesus, and that can be an authentic experience. By arguing that freewill is an illusion that doesn't exist, Harris seems to be excusing murderers and other criminals for their crimes. In his book *Free Will,* Harris even argued that if we could change places with the rapist murderer Steven Hayes, we would have no choice but to become rapists and murderers like Hayes, because we would have his genetics and his life experiences. The combination of Hayes's genetics and living in his same environment would make it impossible for us to react any differently. We would kidnap, rape, and murder when the opportunity presented itself and have no choice in the matter or ability to act otherwise. If that is true, how could we justify putting people in prison for committing violent crimes? Doesn't that conclusion absolve criminals from any responsibility for their actions? How do we justify punishing the guilty if they truly have no choice except to commit their crimes?

Wouldn't the absence of freewill make us slaves to our own DNA and the environment around us?

Compatibilism

Free will and determinism are often seen as being mutually exclusive, or incompatible with each other. Either one or the other is believed to be true. If one is true, the logical conclusion

---

[141] Harris, Sam. "Sam Harris: The Self is an Illusion | Big Think", Big Think, September 16, 2014, https://www.youtube.com/watch?v=fajfkO_Xolo

must be that the other is false. However, compatibilism is the belief that both free will and determinism can be true. Our environment tends to limit the choices we can make, and our DNA dictates whether our bodies are predisposed to illness or disease.

Biologist (and atheist) Jerry Coyne offered a compatibilism quiz on his blog, Why Evolution is True. There were four questions. My attempt to answer these questions follows:

1. What is the definition of free will? **The ability to make a moral choice between right and wrong.**

2. What is free about it? **Free will is a gift from God. Gifts are received freely.**

3. Do other species have free will? Do computers? **I wouldn't claim to know whether animals have free will, but I do know that computers do not. Humans created computers, and humans don't have the power to grant their creation free will.**

4. Why is it important to have free will versus agency? What new knowledge does your concept add beyond reassuring people that we have free will? **Agency would make a third party responsible (or to blame) for making choices. Free will means we can make choices for ourselves.**

I wasn't sure I understood the second part of the fourth and final question, so I only responded to the first part. I'm not sure that my concept of free will adds any new knowledge to the world.

Corroborated Veridical NDE Perceptions

The Near-Death Experience is a unique and interesting phenomenon during which a person in the process of almost dying often experiences the sensation of temporarily separating from their physical body and viewing their body from a third-person perspective. They claim to see deceased loved ones and often to experience a life review. In some of these cases, the dying person claims to learn new information while their brain is temporarily incapacitated, and that information can later be independently investigated and verified to be accurate and true. One of the more interesting examples of corroborated veridical NDE perceptions comes from the account of Michaela Roser, who had been involved in a very serious automobile accident and was undergoing emergency surgery. Michaela's family members had gathered in the hospital cafeteria to wait for information about the surgery's success, and Michaela claimed to have seen them in the cafeteria while her physical body remained in surgery. After she finally awoke following the surgery, Micheala reported witnessing a specific conversation in which both her maternal and paternal grandmothers announced they would smoke a cigarette with her father. Micheala's father was a smoker, but both grandparents did not smoke, so Michaela found the situation quite amusing. However, her mother was quite shocked to learn that Michaela had accurately recalled specific details of a real conversation that had taken place in the hospital cafeteria while at the same time her physical body had been in surgery in a completely different wing of the hospital, in the operating room.

The Near-Death Experience might be dismissed as theistic wishful thinking if it weren't for corroborated veridical NDE perceptions that provide evidence to which the scientific method may be applied. The subject learns new information that can be independently investigated and confirmed to be true, information the subject allegedly learned under duress that otherwise they should not know.

For a long time, my go-to example of corroborated veridical NDE experiences has been the NDE account of Pam Reynolds, which occurred while she was undergoing surgery for a brain aneurysm. While Pam was heavily sedated, with her eyes taped shut and her ears plugged with noise-disrupting ear plugs, she (allegedly) overheard a conversation between a cardio-vascular surgeon and the neurosurgeon leading the team and recalled the details after waking up in recovery. However, Dr. Gary Habermas, from whom I first learned about the corroborated veridical NDE phenomena, said Pam Reynolds was not his "gold standard" for the NDE experience, so I chose to use Michaela Roser instead because the details of her NDE are equally as impressive as Pam's case.

Biochemist and parapsychological researcher Rupert Sheldrake said:

> *"If the whole universe is made of unconscious matter, how come we're conscious? We ought not to be, if everything's unconscious. And some philosophers say, well, the brain does all these things physically, and consciousness is an epiphenomenon. It doesn't do anything. Others say that consciousness is an illusion produced by the brain, but then their critics point out that doesn't really answer the question, since illusion is a form of consciousness. So, you're simply explaining consciousness in terms of consciousness. That's why it's called the hard problem. There's been very little progress in the philosophy of mind about this problem until recently with the development of panpsychism. And panpsychism is primarily fashionable within philosophy of mind, surprisingly fashionable, even though it's very similar to ancient animalistic worldviews we're all supposed to have grown out of."*[142]

Unsurprisingly, Sam Harris has a different opinion:

---

142 Sheldrake Rupert. "A 90 Second Explanation of Panpsychism — Rupert Sheldrake", Institute of Art and Ideas, Jan 15, 2025, https://www.youtube.com/shorts/bKA4ezv7pM4

> *"This picture of a soul that will survive death is a picture of an intelligence that's losing none of those capacities. It's just kind of witnessing the loss of those capacities, and then we'll float off the brain at death, and we know that's not the way minds work. I mean, anyone who has lost any capacity knows that's not a matter of you being silently, perfectly intact, wondering why the body's not working. No, you've lost this thing. If you can't remember, just think of what it's like not to remember somebody's name. Right? If you're trying to think of that person's name, you know you know their name, but you can't—it's the tip of the tongue phenomenon. It's not like your soul is in there with the name, saying, why won't he say the name? No, you, the subject, don't know the name."*[143]

Then how does Mr. Harris explain terminal lucidity, which is a well-documented phenomenon? People suddenly regain function and abilities as they come closer to death. It is hubris to suggest we know much, if anything, about how the brain really works, and arrogantly flippant to use the words "mind" and "brain" interchangeably. If the mind exists, the soul exists. Otherwise, we only have hearts and brains.

So, what is consciousness? Why, it's a miracle!

Planned Versus Unplanned

If God exists, consciousness was planned. It would be reasonable to assume consciousness is a gift, just as life itself is a gift. It may be difficult to understand and even more difficult to explain, but that just means it is more important to cherish the experience rather than try to analyze it. Conversely, if God

---

143 Harris, Sam. "NO PROOF of a SOUL—Sam Harris and Matt Dillahunty", @PangBurn, June 5, 2023, https://www.youtube.com/watch?v=BTAuW3QhoGM

does NOT exist, then consciousness was unplanned. If consciousness were unplanned, then the best explanation for consciousness would be chemical reactions in the brain. If that is the case, why should we trust our own thoughts? If the universe is unplanned, then all the accumulated improbabilities from the origin of the universe through the origin of intelligence and consciousness should be resolvable. In fact, the improbabilities compound until they become a virtual impossibility.

A virtual impossibility that can logically be resolved by the introduction of a plan.

CONCLUSION

Consciousness is defined as having an awareness of one's environment and one's own existence and thoughts. Conscience is defined as having an awareness of what is right or wrong and the urge to act morally.

The Bible doesn't mention consciousness, but it talks about conscience quite a bit. For example, Hebrews 10:22 says:

> *"Let us draw near to God with a sincere heart and with the full assurance that faith brings, having our hearts sprinkled to cleanse us from a guilty conscience and having our bodies washed with pure water."*

Does Stephen Hayes, infamous for the Cheshire murders, have a guilty conscience? I would say no. He's apparently living his best life in prison as Linda Mai Lee. He blamed his crimes on his (allegedly) suppressed transgender personality. In my opinion, Hayes is simply gaming the system to make the best out of a bad situation. Being a sadistic murderer confined in a men's prison doesn't have nearly as much appeal as a male rapist held in a women's prison.

1 Peter 3:16 says:

> *"Keeping a clear conscience, so that those who speak maliciously against your good behavior in Christ may be ashamed of their slander."*

How can we keep a clear conscience? We can try our best to avoid sin. We're not going to succeed, but if we try our hardest, we won't have as many regrets. Stephen Hayes could be forgiven, but he's not showing any sort of repentance, much less the sincere kind.

1 Timothy 3:9 reads:

> *"They must hold the mystery of the faith with a clear conscience."*

Undoubtedly, there are some Bible verses that confound me even today. I'm not sure if I agree with every specific teaching from my personal faith (Lutheran).

However, I'm reasonably convinced that God exists because it became a logical conclusion when confronted by all the compounding improbabilities that comprise the unplanned universe. The difficulty in believing in the unplanned universe comes from the improbability of the Big Bang, the improbability of cosmic inflation, the improbability of abiogenesis, the improbability of Darwinism, the improbability of intelligence, the improbability of consciousness, and finally, the improbability of the individual, or you.

Romans 1:28-32 says:

> *"And since they did not see fit to acknowledge God, God gave them up to a debased mind to do what ought not to be done. They were filled with all manner of unrighteousness, evil, covetousness, malice. They are full of envy, murder, strife, deceit, maliciousness. They are gossips, slanderers, haters of God, insolent, haughty, boastful, inventors of evil, disobedient to*

> *parents, foolish, faithless, heartless, ruthless. Though they know God's righteous decree that those who practice such things deserve to die, they not only do them but give approval to those who practice them."*

Many modern atheists have convinced themselves sin does not exist and think they do not sin, despite the Bible warning us all have sinned and fall short of the glory of God. The Bible also tells us the wages of sin are death, and not a temporal but eternal death. These atheists have suppressed their conscience. There is no one righteous, not even one. It is by grace we are saved, through faith.

MIRACLE 7
THE ORIGIN OF THE INDIVIDUAL

INTRODUCTION

Our physical body literally wants to heal itself. How remarkable is that? Doctors can give you all the medicines in the world, but it won't make one bit of difference if the body's immune system isn't performing its normal role. Doctors don't cure diseases. They can prescribe medicines that allow our immune system to recover and the body to heal itself, but a doctor can only give us all the right medicines and treatments. If the patient has lost their desire to live, that patient will most likely die.

The seventh and final miracle is the existence of our physical bodies. An orthopedic surgeon who had just performed a delicate repair on one of my broken appendages refused to take any credit when I thanked him for healing me. Instead, he indirectly credited the body's designer by saying, "The body wants to heal itself." By that, he meant my immune system would actually heal the broken bone, and it did. The doctor still deserves some measure of credit for setting the bone in place and immobilizing it with a pin until the healing was complete, but he was also right to suggest my body itself would do the heavy lifting.

Our bodies are organic machines that can, with proper training and conditioning, perform incredible tasks that require unusual strength and agility—and if you've ever seen a silent movie starring Buster Keaton or Harold Lloyd or, more recently, a Jackie Chan movie, you'll know exactly what I mean. In the days before green screens, trick photography, and stunt doubles became popular, Hollywood made movies in which actors often performed death-defying stunts to entertain the audience. However, these same human bodies have been known to perform incredible tasks requiring strength and agility *without proper training and conditioning* when under extreme duress. This phenomenon is known as hysterical strength. For example, a grandmother in her late 50s named Angela Carvallo lifted the front end of a Chevrolet Impala approximately four inches off the ground and held the car in the air until two neighbors replaced a collapsed jack and pulled her son from underneath the vehicle. The shock of seeing her child pinned under the heavy vehicle produced a jolt of adrenaline that allowed Ms. Carvallo to perform a task she would never have been able to carry out under ordinary circumstances.

Nor was Ms. Carvallo the only example of hysterical strength: 22-year-old Lauren Kornacki lifted a BMW 525i off her father after the car fell off a jack. Tom Boyle even lifted a Chevrolet Camaro off a trapped cyclist while the traumatized driver remained in the car.[144] This phenomenon may not happen frequently, but it has happened often enough that it has been documented both in scientific papers and newspaper articles.

When I was about fifteen years old, I vividly remember returning home one day and discovering my older sister wasn't there. It was in the middle of summer, so it wasn't completely out of character for my sister to not be home in the middle of the afternoon. In fact, it might normally have been expected that she would not be home at that time. My sister was quite popular and socially active. We were out of school for the

[144] Huicochea, Alexis. "Man lifts car off pinned cyclist", Arizona Daily Star, July 28, 2006, https://tucson.com/news/local/crime/article_e7f04bbd-309b-5c7e-808d-1907d91517ac.html

summer and only lived about thirty miles from the beach. And it was a beautiful, sunny afternoon, with not a cloud in the sky.

As I stepped through the front door, I called out to my sister, and I felt very unsettled when she did not respond. Even though I had no rational reason to believe there was anything out of the ordinary, I instinctively knew something must be very wrong because, as I was stepping through the doorway and calling her name, somewhere in the recesses of my brain I could hear my sister calling out to me. Even more troubling, it sounded as if she might be in big trouble.

My Aunt Delores “Doty” Clark lived only three houses away, so I hurried over to her house and asked her if she knew where my sister was. Aunt Doty tried to make light of the situation and reassure me that nothing was wrong, but uncharacteristically, I challenged her and demanded to know the truth. I knew something was wrong with my sister, and I also instinctively knew my aunt knew what had happened to her, but didn’t want to tell me for some reason. I stared at my aunt. She knew the truth. I knew that she knew. And now, staring back at me, she knew that I knew she knew. The situation was getting complicated.

So, with some reluctance and trepidation, she told me the truth—my sister had gone to the beach with her friends and a guy sitting nearby had spiked her drink while she wasn’t looking, and she temporarily lost consciousness. My aunt finally relented and told me the truth, that my sister was at the hospital as we were speaking, having her stomach pumped out. My sister recovered quickly, and I soon forgot the story had even happened. Years passed. Then, during a casual conversation with a friend, the memory abruptly returned to my mind after years of being completely forgotten.

Suddenly, I could vividly remember the experience and my emotional state of mind when my aunt tried to convince me nothing was wrong, but somehow, I instinctively knew better. I didn't exactly know how I knew, but I knew something was wrong with my sister when I distinctly heard her voice inside my head, calling for help. I can't explain the experience with rational words because the experience itself was irrational. I became emotional as I relived the memory. Somehow, I had become aware that something was terribly wrong with my sister upon walking in the front door that day.

I've never even discussed the experience with my sister to this day. I have no idea what her memories are of that day. However, I've never forgotten that memory, even though my mind had suppressed it for years. It wasn't until I was having a casual conversation with a friend and some innocuous comment for whatever reason jarred the memory loose from the dark recesses of my mind, and suddenly, I was reliving the moment as if it had happened yesterday. It was as if a powerful tsunami-sized wave of emotion just swept right over me in that moment.

My brain is most certainly not like the hard drive of a computer. The record of that event had not been physically etched into the cells of my brain, written and overwritten numerous times, and suddenly, I was able to retrieve this information that had once been stored and then forgotten. The brain doesn't mimic a computer; a computer attempts to mimic the behavior of a mind. Like a brain, a computer has short-term memory and long-term memory. Short-term memory is stored in RAM (Random Access Memory), but long-term memory is written to the hard drive. Where is the "hard drive" in the human brain that stores our long-term memories? If we were to cut into the brain, we wouldn't be able to see the information stored inside the mind. If we read a hard drive, we can retrieve the zeros and ones and reconstruct the information stored, even if the data has been deleted and written over. We have special programs that perform that function. Where in my brain was that distant memory stored?

In a computer, short-term memory is wiped clean every time the machine is powered on and off. Everything that had been stored in RAM is lost when the computer is shut down. Long-term memory exists until it is deleted, and in reality, it continues to exist there, though hidden from the user, until the disk has been overwritten. But our brains don't literally have our memories physically etched into their cells... or do they? We don't really know, but we do know how to make assumptions.

EVIDENCE FOR THE INDIVIDUAL

Physicist Michio Kaku said:

> *"The two greatest mysteries in all of science, in all of science, the two greatest mysteries are, one, what happened before creation? Why did we have a big bang? What banged? Are there other universes, a multiverse before the big bang? That's outer space. Then the second mystery is inner space. What goes on behind your eyeballs? We have one hundred billion neurons in your brain. As many as stars in the Milky Way galaxy. Each neuron is connected to ten thousand other neurons. So, what is the brain?"*[145]

Why, it's a miracle, Dr. Kaku. Don't believe me? Well, consider this: As recently mentioned, a computer has both short-term memory and long-term memory.

Let's say your computer has four gigabytes (4GB) total of memory. A certain portion of that memory, the long-term memory, is occupied by the computer's operating system. The short-term memory is used by computer applications such as internet browsers, email clients, and applications such as the Microsoft Word program used to write this book. As I write in

---

145 Kaku, Michio. "The Two Greatest Mysteries in Science", The Science Fact, November 16, 2023, https://www.youtube.com/shorts/EC_wGGqxZCM

the book, the characters typed are stored in memory until I click the Save icon, which writes the new information to the computer's hard drive. However, I might type several hundred words, perhaps even a thousand words, and neglect to click the Save button...and the computer freezes or crashes. I didn't have the AutoSave feature enabled, so the computer didn't keep any record of the words I just typed.

So, what do I do? I simply reboot the computer and type the same words in again. Will it be verbatim, exactly what I typed before? Probably not. It might be an improvement.

But the point is, I still remember what I just typed, and I can certainly type it in again. It may not be verbatim, but the same idea will still be communicated. The question is, where is this information stored inside my brain? In the computer, it's easy to say where the information is kept. Information can be manufactured and stored in memory, but for that information to persist beyond the computer's power being cycled, it must be written to the computer's hard drive. Otherwise, it is lost the moment the machine powers down. Question—how does the human mind record long-term memories? We can recall details of events years, even decades, after they occur. How does the brain remember?

Christopher Nolan is one of the most brilliant directors working in Hollywood today, bringing to life fantastic movies such as *The Dark Knight* trilogy, *Inception*, and *Oppenheimer*. Arguably, his best movie was one of his first, a brilliant film titled *Memento*. The plot of *Memento* is unforgettable—Leonard, a man suffering from anterograde amnesia (short-term memory loss), while trying to hunt down his wife's killer.

Leonard tends to forget new information almost as soon as he learns it, so he tattoos clues to the identity of his wife's killer all over his body. The story maintains two different timelines: a color timeline tells the story in reverse, while a black-and-white timeline tells the story from beginning to end from Leonard's perspective. The final result is a cinematic masterpiece that ultimately leaves the audience shocked and bewildered by what they just watched—a wildly entertaining film that explores the question of how the brain forms and retains memories in the midst of solving a murder mystery.

We are told through flashbacks that Leonard accidentally hit his head on a mirror while struggling with the home intruder who murdered his wife, which led to the development of his rare, disabling condition of being unable to form new memories. A blow to the head is the explanation given in the movie...but what is it explaining? A short-term memory might include what you had for breakfast this morning. An example of a long-term memory might be where you were and what you were doing on September 11, 2001, when the Twin Towers fell. How does the brain record short-term and long-term memories? What makes them different? It's easy to say with a computer. It's not as easy to say with a human being.

The human body is far from a simple, self-contained organism. The body is also home to trillions of other organisms, single-celled organisms that live inside our bodies and perform vital functions, such as breaking down food compounds and synthesizing nutrients from the food we consume. Additionally, our bodies are composed of trillions of cells that serve as vital parts that contribute to the existence of the primary organism, which is your physical body.

The human body has 11 major organ systems. These include:

1. **Circulatory System**: Includes the heart and blood vessels.

2. **Respiratory System**: Composed of the lungs and airways.

3. **Digestive System**: Includes the stomach, intestines, and liver.

4. **Nervous System**: Composed of the brain, spinal cord, and nerves.

5. **Muscular System**: All the muscles in the body.

6. **Skeletal System**: Consists of bones and joints.

7. **Endocrine System**: Includes glands that release hormones.

8. **Reproductive System**: Involves organs like the ovaries and testes.

9. **Excretory System**: Composed of the kidneys and bladder.

10. **Immune System / Lymphatic System**: Includes the lymph nodes and spleen.

11. **Integumentary System**: Includes the skin, hair, and nails.

Each of these systems exists to perform a specific function that allows for the existence of the body as a whole. The body needs for all of these systems to function normally for the body to survive in the long term. The circulatory system exists to transport blood, oxygen, and nutrients throughout the body. The respiratory system provides the ability for oxygen intake and carbon dioxide removal. The digestive system contains organs that break down nutrients and help the body absorb them. The central nervous system controls the body's reactions to stimuli and its responses. The muscular system makes movement possible. The skeletal system provides the body's structure and morphological form. The endocrine system controls growth and metabolism. The reproductive system allows the body to produce offspring. The excretory system eliminates waste from the body. The immune system allows the body to heal itself. Finally, the integumentary system protects the body from exposure to viruses and pathogens.[146]

The human body is a perfect example of synergy, where the whole becomes greater than the sum of its parts. Without the continuous function of any one of these eleven crucial systems that exist inside the human body, that person would die.

Complete List of Human Organs in the Body and Their Functions

Here is a comprehensive list of the organs in the human body with their functions: [147]

Central Nervous System:

- Brain: Controls body functions and processes.
- Spinal Cord: Transmits signals between the brain and body.

Circulatory System:

- Heart: Pumps blood throughout the body.
- Blood Vessels (arteries, veins, capillaries): Transport blood.

Respiratory System:

---

146 Helmenstine, Anne. "Organs in the Body: Diagram, List, and Functions." Science Notes, September 2, 2025, https://sciencenotes.org/organs-in-the-body-diagram-list-and-functions/

147 Helmenstine, Anne. "Organs in the Body: Diagram, List, and Functions." Science Notes, September 2, 2025, https://sciencenotes.org/organs-in-the-body-diagram-list-and-functions/

- Lungs: Facilitate oxygen intake and carbon dioxide removal.
- Trachea: Connects the throat to the lungs.
- Bronchi: Passageways for air into the lungs.

Digestive System:

- Stomach: Breaks down food for digestion.
- Small Intestine: Absorbs nutrients from food.
- Large Intestine: Absorbs water and forms feces.
- Liver: Detoxifies chemicals and metabolizes drugs.
- Pancreas: Produces digestive enzymes and hormones like insulin.
- Gallbladder: Stores bile for fat digestion.

Excretory System:

- Kidneys: Filter blood to produce urine.
- Ureters: Transport urine from kidneys to bladder.
- Bladder: Stores urine.
- Urethra: Removes urine from the body.

Endocrine System:

- Pituitary Gland: Regulates other glands and growth.
- Pineal Gland: Produces melatonin for sleep regulation.

- Thyroid Gland: Controls metabolism.
- Parathyroid Glands: Regulate calcium levels.
- Adrenal Glands: Produce hormones like adrenaline.
- Pancreas (dual function): Regulates blood sugar.
- Ovaries/Testes: Produce sex hormones.

Immune and Lymphatic Systems:

- Spleen: Filters blood and supports immunity.
- Lymph Nodes: Filter lymph and produce white blood cells.
- Thymus: Matures T-cells for the immune system.

Reproductive System:

- Ovaries: Produce eggs and hormones.
- Testes: Produce sperm and testosterone.
- Uterus: Supports fetal development.
- Prostate Gland: Produces seminal fluid.

Integumentary System:

- Skin: Protects the body and regulates temperature.
- Hair: Insulates and protects skin.
- Nails: Protect the tips of fingers and toes.

Sensory Organs:

- Eyes: Enable vision.
- Ears: Facilitate hearing and balance.
- Tongue: Detects taste and aids in speech.
- Nose: Detects smell.

Skeletal System:

- Bones (206 total): Provide structure and support.
- Bone Marrow: Produces blood cells.

The list currently omits smaller or less prominent organs that are essential but often overlooked in traditional classifications. Here are examples of organs that could be added to reach a more comprehensive list:

- **Salivary Glands** (Parotid, Sublingual, and Submandibular): Produce saliva for digestion and oral health.
- **Lymphatic Structures** (such as Tonsils and Adenoids): Play roles in immunity and filtering pathogens.
- **Mesentery**: A fold of tissue anchoring the intestines, now recognized as an organ.
- **Adipose Tissue**: Fat stores, recently classified as an endocrine organ due to its hormonal activity.

- **Vas Deferens and Seminal Vesicles**: Male reproductive organs aiding in sperm transport and seminal fluid production.

- **Bartholin's and Skene's Glands**: Female reproductive glands involved in lubrication.

- **Inner Ear Structures** (e.g., Cochlea and Vestibular System): Responsible for hearing and balance.[148]

Truly, the complexity and completeness of the design are remarkable. There appears to have been remarkable attention paid to even the most insignificant detail. The vital organs include the brain, heart, liver, and at least one of the two lungs and one kidney. Technically speaking, we know of cases where most of a person's brain is missing, but that person is still able to function normally, but those examples are rare. Their quality of life may be diminished, but a person can survive without the gall bladder, appendix, spleen, reproductive organs, eyes, and, as noted earlier, one kidney or one lung, given the redundancies.

As the very impressive and amazing work of Ms. Anne Helmenstine suggests, the body is very complex. I would further argue that the body is far too complex to have evolved through billions of small, intermediate steps from less complex organisms. It clearly appears to be the product of an intelligent design. The best argument against intelligent design is that there are flaws or perceived errors in the design. However, I would like to remind everyone that even a poor design is still a design, and the argument suggesting there are flaws in the location of the vas deferens tube or the design of the eye's retina

---

[148] Helmenstine, Anne. "Organs in the Body: Diagram, List, and Functions." Science Notes, September 2, 2025, https://sciencenotes.org/organs-in-the-body-diagram-list-and-functions/

could be due to the limitations of human understanding and intellect.

The Observer Effect

There is a strange phenomenon in physics called the observer effect in which the mere act of observation alters the behavior of the particles or phenomena being observed.[149] This effect has been noted in physics, psychology, and other fields of science, where the presence of an observer changes the behavior or state of the object being observed. Is the observer effect the result or a feature of our anthropic universe, which is a universe fine-tuned for intelligent observers to exist?

The Double Slit Experiment and the Which-Way detector

In 1803, British scientist Thomas Young conducted the first double slit experiment to confirm that Sir Isaac Newton's theory that light transmits as particles. The experiment is incredibly simple: merely by cutting two small slits spaced close together into a piece of cardboard or a similar medium and positioning the double-slitted medium above another flat surface so that light from the source (either sunlight or lasers) will shine through the two slits. The experiment revealed that light travels in waves—multiple bands of light were displayed on the targeted medium in this famous experiment in quantum physics.

The experiment has been repeated many times since, with some clever modifications, and here is where it really gets interesting. A device known as a which-way detector may be added to the experiment to determine which of the two slits the particle of light passed through. The detector may be placed before or after

---

[149] Vedyanathan, Venkatesh. "What is the Observer Effect in Quantum Mechanics?", ScienceABC, May 10, 2019, https://www.scienceabc.com/pure-sciences/observer-effect-quantum-mechanics.html

the double slitted barrier, and the effect will remain the same—when the detector is turned on, the patterns displayed on the targeted medium go from multiple bands of light down to two. However, when the detector is turned off, multiple bands of light are again displayed, implying that the very act of observation (or measurement) affects the outcome of the experiment.

It reminds one of a variation on the old question that has puzzled philosophers for decades: if a tree falls in the forest and there is no one there to hear it, does the tree make a sound? Can a flower be described as beautiful if no one ever sees it? The double slit experiment with a which-way detector implies the answer to the question is that observers *do* make a difference. Merely by measuring which slit a photon passes through changes the output of the experiment. Physicist Brian Cox said:

> *"Let's imagine that in our galaxy of four hundred billion suns there is just us that thinks. There may be microbes all over the place but in terms of things that think and can feel and in a very real sense bring meaning to the universe—all these things we've talked about. The beauty of these galaxies —they're not beautiful if there's nothing there to perceive them, right? They are just galaxies."*[150]

ANALYSIS OF THE INDIVIDUAL

The Split-Brain Experiment

The fundamental problem with conducting split-brain experiments is the lack of viable candidates for test subjects. Only people who have undergone callosotomy surgery to split the corpus callosum as a drastic treatment for epilepsy are

---

150 Cox, Brian. "What If our SUN became a black hole?", @spacein1minute, https://www.youtube.com/shorts/edVqAhwaNNo

eligible for split-brain experiments. A man named Joe underwent a callosotomy to prevent grand mal seizures due to his epilepsy. After the crucial surgery had been performed to alleviate his condition, scientists seized the opportunity to learn how Joe's brain functioned. An experiment was conducted where a screen with two images was shown to Joe. On the left side of the screen was the picture of a hand saw. On the right side of the screen was a hammer. Joe was asked to close his eyes and draw a picture of what he saw. He drew a saw. He is then asked what he'd seen, and he replied "hammer."

And when asked why he had drawn a saw, Joe responded, "I don't know."

During the experiment, Joe had been staring at a computer screen with the two images spaced just far enough apart so that each of his eyes would only see one of the two images, a saw and a hammer. Under normal conditions outside of the experiment, Joe would see the same images or words with both of his eyes, but due to the careful preparations for the experiment, he'd seen the saw and a hammer independently with a different eye. The language center in his brain only saw the hammer, but the visual/artistic side of his brain only perceived the existence of the saw, so he verbally responded "hammer" when he'd actually drawn a saw.

Presumably, anyone and everyone with an intact corpus callosum would have seen both the saw and the hammer. Under normal conditions, Joe would have also seen both. The experiment was designed to test the residual effects of the surgery, and the results were fascinating. Under normal conditions, the experiment would have been impossible, but the necessity for Joe to undergo the surgery provided the opportunity. The left hemisphere of the brain is the language center. It connects to the right side of the body and the right eye. When Joe is shown a picture, he can use words to verbally respond from the right side of his brain. However, when asked to draw a picture of what he's seen, Joe uses his left hand to convey the images seen by his right eye. The left eye feeds information to the right hemisphere of the brain, and the right eye feeds information to the left hemisphere.

The brain's information processing hub is the corpus callosum. Only when the corpus callosum is intact does the brain recognize that two images are displayed on the screen: a saw and a hammer. With a split-brain patient, the eyes still see both images, but the brain processes each image independently. Depending on which hemisphere of the brain was queried, Joe simply responded to what each individual eye saw. This experiment demonstrated that different areas of the brain process different types of information. One side of the brain is responsible for processing visual information, and the other side of the brain is responsible for verbal communication. When the corpus callosum is intact, the brain functions as one fully integrated unit, but when it is divided by surgery, the two halves of the brain can continue to function independently.

However, Justine Sergent devised a split-brain experiment in which participants were shown two arrows on a divided screen. She asked the participants if the arrows pointed in the same direction, and every time their answer was correct. The participant was always able to say whether the arrows pointed in the same direction, even though the images were seen independently by the two halves of their divided brain. Sergent concluded there was something in the mind that wasn't in the physical brain. Researcher Yair Pinto has conducted experiments where split-brained patients are independently shown two parts of a story (for example, a baseball and a broken window) and the patient is nevertheless able to construct a coherent narrative from the two elements of the story (the baseball broke the window) even though no part of the brain saw both the baseball and the broken window. Again, the mind is greater than the sum of its physical parts.[151]

Wilder Penfield conducted a series of groundbreaking experiments while performing neurosurgery on conscious patients in order to map the various areas of the brain and pinpoint the location where seizures occurred. Penfield successfully located a movement center in the brain, as well as a sensation center, a visual center, a hearing center, emotional centers, and memory centers, but no center for abstract thought. He found no center for reason.

Thanks to the research efforts of people like Wilder Penfield and Justine Sergent, we now know quite a bit about the human brain. We know what various areas of the brain do and how specific areas in each hemisphere operate in conjunction with other areas across the brain in order to provide a cohesive and coherent interpretation of stimuli provided by the world around us. We know that when a corpus callosotomy is performed on patients with severe epilepsy, the two halves of the brain can independently react to different stimuli, yet the mind remarkably manages to construct a coherent narrative by piecing together two different puzzle pieces independently observed by the separated halves of their brain.

Planned Versus Unplanned

If they are still alive, you may be able to ask your parents whether your birth was planned or unplanned. And if your birthday is less than nine months after their wedding anniversary, you already have your answer—you were unplanned. All joking aside, even if your conception was not intended by your parents, the universe itself clearly appears to have been planned, and thus everything within the universe

[151] Egnor, Michael. "Is the Soul Real? Neuroscience's Surprising Answer", Discovery Science, January 11, 2026, https://www.youtube.com/watch?v=Vr3hbrPAmrk

also must have been planned. To believe in an unplanned universe, we must believe order came from chaos and the universe was created from nothing for no reason. That belief requires more faith than I can muster. If the universe is planned, God or some other supreme intellect must obviously be the planner.

CONCLUSION

Several years ago, I was taking my dogs for a walk when I somehow broke a bone in my finger. It hurt constantly for several days, and I scheduled an appointment to consult with a skilled orthopedic surgeon about my problem. He suggested putting a pin in my finger to allow my knuckle to heal properly. I countered with the idea of "Ronnie Lott" surgery simply to amputate the finger. It was just a pinkie finger...how bad did I need it?

The same orthopedic surgeon who told me my body wanted to heal itself replied that I get about half of the strength in that hand from the bottom two fingers, so the loss of my pinkie finger would be a significant loss. Furthermore, it was the pinkie finger on my dominant hand, so I would have been crippling myself for the rest of my life for short-term gain. He persuaded me to have the surgery and endure the rehabilitation process that followed for the sake of my long-term benefit, and I'm thankful I did. That same doctor was the one to give most of the credit for my healing to my skeletal system rather than taking it for himself. It sharply brought into focus who and what had really healed me.

My body had truly done almost all the heavy lifting. The doctor stabilized my broken knuckle joint, and his efforts were critical to the healing process, but my own body and my skeletal system literally healed the break. My immune system assisted by recruiting cells that are essential to the healing process. My circulatory system delivered blood and nutrients to the site of the fracture that are crucial to the healing process, and new blood vessels formed while the bone healed. Hormones that regulate calcium levels and bone metabolism were produced by my endocrine system, and my nervous system produced feedback in the form of pain on those occasions I pushed the limits during rehabilitation. Not only did my body heal itself, but different parts of my body had worked in a coordinated effort to facilitate healing. My conscious brain didn't even have to lift a finger (except during physical therapy).

Ronnie Lott, the Hall of Fame San Francisco 49er safety, chose to have part of his left pinkie finger amputated instead of having a similar surgery to mine because he wanted to continue playing football that season. It was a long-term sacrifice for a short—term gain. It seems foolish to forever limit your future potential by amputation just to play in a football game, even if you are a professional football player. Does Ronnie Lott miss the appendage he sacrificed for his football team? Would he have made the same decision if it had been the pinkie finger on his dominant (right) hand? Does Lott regret the decision? During an interview, he joked that his amputated finger looked like ET's head and said, "I could have all of (49ers team owner) Eddie DeBartolo's corporations, and it isn't going to buy me a new finger." One thing is certain—when Ronnie Lott puts on his four Superbowl rings, he won't be wearing one of them on the pinky finger of his left hand.

Naturally, the Bible has several things to say about the miracle of the individual.

Genesis 1:27 says:

> *"So God created mankind in his own image, in the image of God he created them; male and female he created them."*

God said all the way back in the book of Genesis that we were made in His image. Given the fact God doesn't have a physical body, that must mean our intelligence and conscience have been modeled after His mind.

1 Peter 2:9 says:

> *"But you are a chosen people, a royal priesthood, a holy nation, God's special possession, that you may declare the praises of him who called you out of darkness into his wonderful light."*

God's special possession—doesn't that sound wonderful? Some might argue that possession sounds a lot like slavery, but I would remind them that God has also given us the gift of free will, which means we will always have the choice whether to worship Him or worship ourselves.

Ephesians 2:10 says:

> *"For we are God's handiwork, created in Christ Jesus to do good works, which God prepared in advance for us to do."*

Romans 5:8 reads:

> *"But God demonstrates his own love for us in this: While we were still sinners, Christ died for us."*

Later in Romans Chapter 12 verses 4 and 5, Paul literally compares the church to the human body, writing,

> *"Now we have as many parts as one body, and all the parts do not have the same function in the same way we who are many are one body in Christ and individually members of one another."*

Paul is making an analogy comparing the functions of the

human body to the function of individuals within the Christian church, which is sometimes called "the body of Christ." Just as the liver, kidneys, eyes, heart, and lungs all perform separate and unique functions within the human body, Paul points out that the church functions in a similar manner. Some members of the church are better suited to teaching, while others preach or minister to those in need.

Why was Jesus forced to die on the cross for our sins? Was there a purpose for the crucifixion? If Jesus had to die, he could have died in numerous ways. He could have been burned to death or drowned. He could have been beheaded like John the Baptist. Why crucifixion? Apparently, crucifixion was an especially brutal, agonizing way to die. Death by immolation is usually over in a matter of seconds. Beheading is nearly painless; the suffering is typically over in a fraction of a second unless the process is botched somehow.

In contrast, crucifixion lasted for hours. It wasn't merely death; it was torture. The Romans were pretty good about keeping records, so we'd know if there had ever been an instance where the Romans intended to kill someone through crucifixion and tried but failed to complete the job. Muslims will sometimes argue that an imposter was crucified in lieu of Jesus, but there is no evidence indicating the Romans crucified the wrong guy. If Jesus died on the cross and was resurrected from the dead, then he was exactly who he claimed to be.

1 Corinthians 6:19 says:

> *"Do you not know that your bodies are temples of the Holy Spirit, who is in you, whom you have received from God? You are not your own."*

We come from dust, and to dust we shall return. We only have temporary ownership of these physical bodies. Our minds (or souls) are the only part of us with the potential to survive physical death of these mortal bodies.

1 Corinthians 12:12 says:

> *"For just as the body is one and has many parts, but all its many parts form one body, so it is with Christ."*

The men (and possibly women) who wrote the Bible are often referred to as illiterate shepherds by atheist critics, but in the verse from 1st Corinthians, they are communicating basic (and accurate) medical knowledge. The body is indeed composed of many parts, many of which were identified earlier in the chapter.

But two verses in particular establish how long God has known us.

Jeremiah 1:5 says:

> *"Before I formed you in the womb I knew you, before you were born, I set you apart; I appointed you as a prophet to the nations."*

And a verse found in Psalm 139 reads:

> *"For you created my inmost being, you knit me together in my mother's womb."*

Both of those verses claim that God has known us since our conception.

Ephesians 4:16 says:

> *"From him the whole body, joined and held together by every supporting ligament, grows and builds itself up in love, as each part does its work."*

In other words, just as the systems inside our physical bodies collaborate to perform work, our spirit grows as our love grows.

The Wistar symposium was a famous meeting in which a group of experts in mathematics and computers debated a group of biologists led by Ernst Mayr.

Dr. Murray Eden had set off fireworks when he delivered a scintillating lecture on the mathematical improbability of Darwinism working in a real-life environment, and a number of passionate exchanges followed as tempers began to flare. At one point, botanist Conrad Zirkle addressed symposium chairman Sir Peter Medawar and quite sarcastically said:

> *"Mr. Chairman, I wish merely to indulge in a little improbability, one that is at least as great as that cited by Dr. Eden (a mathematician critical of Darwinism). If we can assume, I think quite reasonably, that our parents were heterozygous for about 10,000 loci, we can see how slight the chances are that any one of us would have been born instead of some non-existing brother or sister. The number of our ancestors also increases exponentially per generation back to a point where everyone probably is descended from everyone but of course, in a different degree. Now what is the probability of any one of us being here in this room after the human race has been here on Earth for about 1 million years? I am convinced the chances against any one of us having been born is practically infinite, and this forces me to accept solipsism and to assume that this room is empty, except for myself, and the only existence any of you have is in my imagination."* [152]

---

[152] Moorhead, Paul S. and Kaplan, Martin M. Mathematical Challenges to the Neo-Darwinian Interpretation of Evolution. Page 53. New York. Alan A. Liss, Inc. Print.

It sounds like Dr. Zirkle had become frustrated that the mathematicians kept asking tough questions that the biologists couldn't answer except to say their theory must be correct, because the only true alternative was special creation by a supernatural God. I'm quite happy to reduce the problem down to two basic probabilities to compare: in a planned universe, life was created by God. Evolution (mutation and adaptation) may account for diversity within a kind of animal (cat, dog, bear, human, fish), but does not establish any sort of connection between different kinds like fish and whales. Modern evolution theory teaches universal common descent, meaning every living organism on Earth, whether plant, animal, fish, bird, plus every other creature, is directly related due to descent with modifications because of sexual reproduction that occurs over vast amounts of Time. The belief might be theoretically possible, but its unlikelihood is exacerbated by all the highly improbable events that need to succeed before evolution can occur. Life cannot evolve until it exists, and thus the probability of evolution is negatively influenced by the improbability of the origin of life from inanimate matter, whether by spontaneous generation, abiogenesis, or some other undirected means. Our lives literally depend on the success of seven independent miracles, and without any one of them, you would not be here to read this sentence.

Assuming you did read it, you should thank God with all your heart.

FINAL THOUGHTS

Many atheists are very intelligent people. Their intelligence often proves to be detrimental because they tend to assume they are the smartest people in the room. Everyone has a cognitive bias either toward or against belief in God. The stereotypical atheist might say something such as the Bible was written by ancient people who didn't understand modern science, which is true. Science didn't exist when the Bible was written—yet these people who wrote the Bible, often mocked and ridiculed as being illiterate sheepherders (in spite of the fact illiterate people can neither write nor read), somehow got the basic order of creation correct. Lucky guess?

The universe and everything in the universe exist either because an intelligent mind planned for this perfectly fine-tuned universe and flawlessly executed the plan OR because an unplanned universe with living organisms resulted from a series of extremely fortuitous events that might even be described as a confluence of happy accidents. In a purely materialistic world, the spirit does not exist, but we know the soul, or life force, does exist, and it literally animates matter. Therefore, the purely materialistic worldview must be false. In the spiritual worldview, the life force that animates our physical bodies is called a soul, but the soul is the same thing as a spirit. God is pure spirit without a physical body.

One lucky event is theoretically possible even if it doesn't appear to be reasonable or plausible. However, the probability problem for an unplanned universe gets progressively worse with each new improbability. The improbabilities compound because all seven miraculous events absolutely must have a successful outcome in order for us to exist, and we do exist. Beginning with the Big Bang (or cosmic inflation), the origin of life, the diversity of life, the origin of intelligence, the origin of consciousness, and even the origin of the individual all qualify as miracles. Unlikely events such as winning a lottery do occur in spite of a very low probability of success, but who should expect to win seven consecutive lotteries, no matter how low the odds of success might be? Seven consecutive winning lotteries, with no losses? I believe in the possibility of luck, but the luck necessary for all seven miracles to occur in the necessary order for us to exist defies logic and reason.

An unplanned Big Bang is an extremely improbable event. Unplanned and undirected cosmic inflation is another extremely improbable event. Both improbabilities must be resolved in order for the universe to exist. This (allegedly) fine-tuned universe is essential for the origin of life. Life cannot evolve until it exists, which means evolution absolutely depends on the success of abiogenesis. Life, and not just human life, is intelligent. Humans are also conscious, but it is unknown whether other life is conscious. Every living organism, plant and animal alike, is a kind of miracle because it has DNA that makes it uniquely identifiable from every other specimen of its kind. Even insects. Even trees. That in itself is a miracle.

The cumulative effect of seven consecutive improbabilities adds up to a virtual impossibility. The only reason one should reject the idea of a planned universe is to reject the idea of the planner responsible. The precision and execution of everything necessary to produce order from chaos. The best explanation for the existence of intelligence is superior intelligence; otherwise, we should assume that the smartest human is the highest form of intelligence that exists.

You are the culmination of the seven miracles performed by God that led to the creation of your physical body. You should consider expressing gratitude to your Creator for this wonderful gift you've been given.

ABOUT THE AUTHOR

My identity should be irrelevant to the reader of this book, truth be told. Most of "my" evidence for divine creation and intelligent design actually comes from the quoted words of physicists, biochemists, paleontologists, and biologists—the foremost experts in their respective fields. My contribution primarily consisted of compiling information gleaned from reading the work of others, not making any bold or new scientific discoveries of my own.

I'm not a scientist. Nor do I claim to play one on television. All of the credit for the scientific research belongs to those scientists and intellectual giants on whose shoulders I stand. The blame for any flaws in my logic, conclusions, and opinions belongs only to me.

My college matriculation culminated with a bachelor's degree in business administration from the University of Georgia, majoring in Management Information Systems. Basically, that's a fancy way of saying that my collegiate career ended when I received an undergraduate computer science degree from the business school.

After graduation, I spent almost two decades writing computer software. Early in my career, I displayed such dedication to the job and tenacity for problem solving that I was selected from a large team of developers to travel the world as part of an elite "fly & fix" team for international product support, taking me to strange and exotic locations like Australia, Singapore, Hong Kong, and Paramus, New Jersey.

After twenty years of software development and brief tenure as a real estate investor, my journey to becoming a writer literally began by accident. The television had been left playing in the background while I worked on developing a business plan for a friend of mine, when a rather famous atheist and prominent scientist said something that caught me off-guard and so completely contradicted my world view that I spent the next several years voraciously consuming every book I could find about the science he claimed proved that God did not exist. Since I was equally confident a creator God was somehow responsible for my origin, obviously we could not both be right. My research became my first nonfiction book.

Once I began writing, I submitted several articles to the online publication American Thinker, which led to my being interviewed live by Editor-in-Chief Thomas Lifson on The Dennis Miller Radio Show.

After *Divine Evolution* was released by Epress-Online, I spent several more years of research as the Atlanta Creationism Examiner, producing a column from 2010 to 2012 that culminated in the writing of this book. During this period, I also enjoyed writing a non-fiction collection of short stories about animal rescue and fostering dogs entitled *Always a Next One*.

My other books are detective novels, written as Rocky Leonard. I mention them because *Coastal Empire*, *Secondhand Sight*, *Premonition*, *Hunter's Omen*, and *Atheist's Prayer* are published novels and you should buy them from your favorite bookseller. I believe that writing detective novels has sharpened my analytical skills and problem solving ability, so it's sort of relevant to mention them.

Prominent atheist and author Ed Buckner, president emeritus of American Atheists, decided my articles were provocative enough that he joined me in a public debate. Also, my friend Rev. Bill Wassner referenced a few of my articles in a course he taught at DePaul University. I am grateful to both Ed and Bill, for considering me a worthy adversary in debate, and recommending a couple my articles to a class of students.

My most recent publication prior to *Seven Miracles* was *The God Conclusion*, which is the most widely read of my works to date. I'm quite proud of the following it has gathered. Visit our Facebook page for lively debate and commentary.

As I wrap up this latest book, I'm pleased to announce that my wife and I have officially retired and relocated from decades of big city life to a quiet small town. Our days are now filled with grandkids, golf, and great weather (not necessarily in that order). Yes, I still make time for writing and probably always will. Thanks for reading.

ABOUT THE COVER

The cover we selected features a bright light surrounded by seven symbols that represent the seven miracles.

The topmost symbol is an atom, representing the creation of the universe. Moving clockwise, the next symbol is a whirling cosmos, representing the expansion of the early universe. Next is a symbol featuring water, essential to the origin of life. The tree of life is featured in the fourth symbol to represent the diversity of life. The fifth symbol is an all-seeing eye, reflecting a sign of intelligence. The da Vinci-like human featured on the sixth symbol is what we chose to suggest consciousness. The seventh and final symbol is the Celtic cryptic known as the triskelion, which represents the unity of body, mind, and spirit.

Finally, the bright light in the center of the cover imagery represents the light of God.

It is very simple, but quite meaningful.

www.ingramcontent.com/pod-product-compliance
Lightning Source LLC
LaVergne TN
LVHW010603100826
845148LV00014B/2833

* 9 7 9 8 9 8 6 3 1 0 2 5 1 *